133
Structure and Bonding

Series Editor: D. M. P. Mingos

Editorial Board:

P. Day · X. Duan · L. H. Gade · T. J. Meyer
G. Parkin · J.-P. Sauvage

Structure and Bonding

Series Editor: D. M. P. Mingos

Recently Published and Forthcoming Volumes

Controlled Assembly and Modification of Inorganic Systems
Volume Editor: Wu, X.- T.
Vol. 133, 2009

Molecular Networks
Volume Editor: Hosseini, M. W.
Vol. 132, 2009

Molecular Thermodynamics of Complex Systems
Volume Editors: Lu, X., Hu, Y.
Vol. 131, 2009

Contemporary Metal Boron Chemistry I
Volume Editors: Marder, T. B., Lin, Z.
Vol. 130, 2008

Recognition of Anions
Volume Editor: Vilar, R.
Vol. 129, 2008

Liquid Crystalline Functional Assemblies and Their Supramolecular Structures
Volume Editor: Kato, T.
Vol. 128, 2008

Organometallic and Coordination Chemistry of the Actinides
Volume Editor: Albrecht-Schmitt, T. E.
Vol. 127, 2008

Halogen Bonding
Fundamentals and Applications
Volume Editors: Metrangolo, P., Resnati, G.
Vol. 126, 2008

High Energy Density Materials
Volume Editor: Klapötke, T. H.
Vol. 125, 2007

Ferro- and Antiferroelectricity
Volume Editors: Dalal, N. S.,
Bussmann-Holder, A.
Vol. 124, 2007
Photofunctional Transition Metal Complexes
Volume Editor: V. W. W. Yam
Vol. 123, 2007

Single-Molecule Magnets and Related Phenomena
Volume Editor: Winpenny, R.
Vol. 122, 2006

Non-Covalent Multi-Porphyrin Assemblies
Synthesis and Properties
Volume Editor: Alessio, E.
Vol. 121, 2006

Recent Developments in Mercury Science
Volume Editor: Atwood, David A.
Vol. 120, 2006

Layered Double Hydroxides
Volume Editors: Duan, X., Evans, D. G.
Vol. 119, 2005

Semiconductor Nanocrystals and Silicate Nanoparticles
Volume Editors: Peng, X., Mingos, D. M. P.
Vol. 118, 2005

Magnetic Functions Beyond the Spin-Hamiltonian
Volume Editor: Mingos, D. M. P.
Vol. 117, 2005

Intermolecular Forces and Clusters II
Volume Editor: Wales, D. J.
Vol. 116, 2005

Controlled Assembly and Modification of Inorganic Systems

Volume Editor: Xin-Tao Wu

With contributions by

R.G. Cavell · L. Chen · Z.-N. Chen · J.D. Corbett ·
F.-R. Dai · A.P. Grosvenor · S.-M. Hu · J. Jiang ·
R. Li · Q. Lin · A. Mar · Z. Ni · T.-L. Sheng ·
X.-T. Wu · S.-C. Xiang

Prof. Dr. Xin-Tao Wu
Chinese Academy of Sciences
Fujian Institute of Research
on the Structure of Matter (FIRSM)
350002 Fuzhou, Fujian
China, People's Republic
xtwu@fjirsm.ac.cn

ISSN 0081-5993   e-ISSN 1616-8550
ISBN 978-3-642-01561-8   e-ISBN 978-3-642-01562-5
DOI 10.1007/978-3-642-01562-5
Springer Dordrecht Heidelberg London New York

Library of Congress Control Number: 2009926183

© Springer-Verlag Berlin Heidelberg 2009
This work is subject to copyright. All rights are reserved, whether the whole or part of the material is concerned, specifically the rights of translation, reprinting, reuse of illustrations, recitation, broadcasting, reproduction on microfilm or in any other way, and storage in data banks. Duplication of this publication or parts thereof is permitted only under the provisions of the German Copyright Law of September 9, 1965, in its current version, and permission for use must always be obtained from Springer. Violations are liable to prosecution under the German Copyright Law.
The use of general descriptive names, registered names, trademarks, etc. in this publication does not imply, even in the absence of a specific statement, that such names are exempt from the relevant protective laws and regulations and therefore free for general use.

Cover design: KünkelLopka GmbH, Heidelberg, Germany

Printed on acid-free paper

Springer is part of Springer Science+Business Media (www.springer.com)

Series Editor

Prof. D. Michael P. Mingos

Principal
St. Edmund Hall
Oxford OX1 4AR, UK
michael.mingos@st-edmund-hall.oxford.ac.uk

Volume Editor

Prof. Dr. Xin-Tao Wu

Chinese Academy of Sciences
Fujian Institute of Research
on the Structure of Matter (FJIRSM)
350002 Fuzhou, Fujian
China, People's Republic
xtwu@fjirsm.ac.cn

Editorial Board

Prof. Peter Day

Director and Fullerian Professor
of Chemistry
The Royal Institution of Great Britain
21 Albermarle Street
London W1X 4BS, UK
pday@ri.ac.uk

Prof. Xue Duan

Director
State Key Laboratory
of Chemical Resource Engineering
Beijing University of Chemical Technology
15 Bei San Huan Dong Lu
Beijing 100029, P.R. China
duanx@mail.buct.edu.cn

Prof. Lutz H. Gade

Anorganisch-Chemisches Institut
Universität Heidelberg
Im Neuenheimer Feld 270
69120 Heidelberg, Germany
lutz.gade@uni-hd.de

Prof. Thomas J. Meyer

Department of Chemistry
Campus Box 3290
Venable and Kenan Laboratories
The University of North Carolina
and Chapel Hill
Chapel Hill, NC 27599-3290, USA
tjmeyer@unc.edu

Prof. Gerard Parkin

Department of Chemistry (Box 3115)
Columbia University
3000 Broadway
New York, New York 10027, USA
parkin@columbia.edu

Prof. Jean-Pierre Sauvage

Faculté de Chimie
Laboratoires de Chimie
Organo-Minérale
Université Louis Pasteur
4, rue Blaise Pascal
67070 Strasbourg Cedex, France
sauvage@chimie.u-strasbg.fr

Structure and Bonding
Also Available Electronically

Structure and Bonding is included in Springer's eBook package *Chemistry and Materials Science*. If a library does not opt for the whole package the book series may be bought on a subscription basis. Also, all back volumes are available electronically.

For all customers who have a standing order to the print version of *Structure and Bonding*, we offer the electronic version via SpringerLink free of charge.

If you do not have access, you can still view the table of contents of each volume and the abstract of each article by going to the SpringerLink homepage, clicking on "Chemistry and Materials Science," under Subject Collection, then "Book Series," under Content Type and finally by selecting *Structure and Bonding*.

You will find information about the

– Editorial Board
– Aims and Scope
– Instructions for Authors
– Sample Contribution

at springer.com using the search function by typing in *Structure and Bonding*.

Color figures are published in full color in the electronic version on SpringerLink.

Aims and Scope

The series *Structure and Bonding* publishes critical reviews on topics of research concerned with chemical structure and bonding. The scope of the series spans the entire Periodic Table and addresses structure and bonding issues associated with all of the elements. It also focuses attention on new and developing areas of modern structural and theoretical chemistry such as nanostructures, molecular electronics, designed molecular solids, surfaces, metal clusters and supramolecular structures. Physical and spectroscopic techniques used to determine, examine and model structures fall within the purview of *Structure and Bonding* to the extent that the focus is on the scientific results obtained and not on specialist information concerning the techniques themselves. Issues associated with the development of bonding models and generalizations that illuminate the reactivity pathways and rates of chemical processes are also relevant.

The individual volumes in the series are thematic. The goal of each volume is to give the reader, whether at a university or in industry, a comprehensive overview of an area where new insights are emerging that are of interest to a larger scientific audience. Thus each review within the volume critically surveys one aspect of that topic and places it within the context of the volume as a whole. The most significant developments of the last 5 to 10 years should be presented using selected examples to illustrate the principles discussed. A description of the physical basis of the experimental techniques that have been used to provide the primary data may also be appropriate, if it has not been covered in detail elsewhere. The coverage need not be exhaustive in data, but should rather be conceptual, concentrating on the new principles being developed that will allow the reader, who is not a specialist in the area covered, to understand the data presented. Discussion of possible future research directions in the area is welcomed.

Review articles for the individual volumes are invited by the volume editors.

In references *Structure and Bonding* is abbreviated *Struct Bond* and is cited as a journal.

Impact Factor in 2007: 4.041; Section "Chemistry, Inorganic & Nuclear": Rank 5 of 43; Section "Chemistry, Physical": Rank 20 of 110

Dedicated to
Fujian Institute of Research on the Structure of Matter,
Chinese Academy of Sciences,
on the occasion of its 50th anniversary

Preface

From the viewpoint of structural chemistry, structure and bonding lie at the heart of rational syntheses that have already contributed to many significant scientific advances in inorganic chemistry and material chemistry, and especially to the discovery of some functional materials. Naturally the first step to novel functional material is "synthesis", and in many cases exploratory synthesis seems to be the only workable route to new compound. However, rational synthesis will surely make property-oriented exploration more fruitful and pleasing.

Success under the guidance of electronic structural features, bonding interactions, chemical reactivity of building units, etc. has been achieved in many systems. We have presented some significant advances on five topics via review-type chapters that were written by five of the leading authorities in their fields. These chapters concern chemical approach to new quasicrystals, discovery of complicated compounds of pnicogen, the tuning of redox levels and oligomerization of triruthenium-acetate clusters, structural modification of monomeric phthalocyanines, and the controlled assembly of amino lanthanide metal-organic frameworks (MOFs).

This volume has shown that the controlled assembly and modification of inorganic systems are accessible and efforts along the way will contribute greatly to the discovery of new functional materials as well as the satisfaction of the curiosity of fundamental research.

Fuzhou, March 2009 *Xin-Tao Wu*

Acknowledgements

I am grateful for Prof. D.M. P. Mingos for his invitation to be a volume editor and his suggestion for the title of this volume. I am also indebted to all the authors who unselfishly spent their precious time in writing contributions for the volume and meeting deadlines. Special thanks are given to Prof. Thomas C. W. Mak in the Chinese University of Hong Kong, Prof. Xue Duan in Beijing University of Chemical Technology, Prof. Bei-Sheng Kang in Sun Yat-Sen University, Guangzhou and Prof. Ling Chen in Fujian Institute of Research on the Structure of Matter, (FIRSM) for their kind help, and Mr. Sheng-Min Hu for his skill and understanding during the proof reading process.

I would like to dedicate this book to FIRSM CAS. It was established by our first Director the late Prof. Jia-Xi Lu and he was also President of CAS.

Contents

**A Chemical Approach to the Discovery of Quasicrystals
and Their Approximant Crystals** 1
Qisheng Lin and John D. Corbett

**Bonding and Electronic Structure of Phosphides,
Arsenides, and Antimonides by X-Ray Photoelectron
and Absorption Spectroscopies** 41
Andrew P. Grosvenor, Ronald G. Cavell, and Arthur Mar

**Oxo-Centered Triruthenium-Acetate Cluster Complexes
Derived from Axial or Bridging Ligand Substitution** 93
Zhong-Ning Chen and Feng-Rong Dai

**New Progress in Monomeric Phthalocyanine Chemistry:
Synthesis, Crystal Structures and Properties** 121
Zhonghai Ni, Renjie Li, and Jianzhuang Jiang

**Controllable Assembly, Structures and Properties
of Lanthanide–Transition Metal–Amino Acid Clusters** 161
Sheng-Chang Xiang, Sheng-Min Hu, Tian-Lu Sheng, Ling Chen,
and Xin-Tao Wu

Index 207

Struct Bond (2009) 133: 1–39
DOI:10.1007/430_2008_11
© Springer-Verlag Berlin Heidelberg 2009
Published online: 5 March 2009

A Chemical Approach to the Discovery of Quasicrystals and Their Approximant Crystals

Qisheng Lin and John D. Corbett

Abstract This review is intended to be a chemist-friendly introduction to what quasicrystals (QCs) and approximant crystals (ACs) are and what chemists may be able to contribute to the field. Readers will first be exposed to a must-know history of QC/ACs, then warmed up with the somewhat distant and prior concepts of metal clusters in halides, oxides etc., and then to polyanionic clusters in Zintl phases and intermetallic systems. Information on these last two has originated over about the last 50 years. We will draw on some more chemical insights and information on how these might be related and applicable to new and expanded QC and AC systems. Then follow our experiences on electronic and chemical tuning of five QC and AC systems and the structural regularities within ACs, from which important clues for quasicrystal structure modeling are evident.

Keywords: Approximant crystals · Electronic tuning · Intermetallics · Pseudogap · Quasicrystals

Contents

1	Introduction	2
2	Traditional Metal Clusters	8
	2.1 Halide, Oxides, and Sulfide Systems	8
	2.2 Zintl Phases	8
	2.3 Polar Intermetallics	9
3	Toward Quasicrystals and Their Approximants	10
	3.1 Polar Intermetallics Containing the Triels	10
	3.2 Chemical Insights from Earlier Works	13
	3.3 A Reward	14
4	Pseudogap Tuning	16
	4.1 Structure and Bonding in $Mg_2Cu_6Ga_5$	16
	4.2 Tuning ACs and QCs from Mg_2Zn_{11}-Type Precursors	18

Q. Lin and J.D. Corbett (✉)

Department of Chemistry and Ames Laboratory, Iowa State University, Ames, IA 50011, USA

e-mail: jcorbett@iastate.edu; qslin@iastate.edu

5	Structure Regularities	27
	5.1 Tsai-Type ACs	27
	5.2 Bergman-Type ACs	32
	5.3 AC Structures and QC Modeling	34
6	Remarks	35
References		37

Abbreviations

AC	Approximant crystal
COHP	Crystal orbital Hamilton population
COOP	Crystal overlap orbital population
DOS	Densities-of-states
e/a	Valence electron count per atom
i-QC	Icosahedral quasiperiodic crystal
LMTO	Linear muffin tin orbital
LRO	Long-range order
nD	n-Dimensional
OR	Oblate rhombohedron
PR	Prolate rhombohedron
QC	Quasiperiodic crystal
SRO	Short-range order

1 Introduction

The discovery of the first Al_6Mn icosahedral quasiperiodic crystal (i-QC) in 1984 [1] broke a traditional serenity within the crystallographic community because this phase exhibited the first clear example of noncrystallographic icosahedral symmetry. This drew scientists from divergent fields into intensive investigations of QCs and their corresponding approximant crystals (ACs) from many viewpoints (theory, synthesis, structure, property, etc.) [2–5]. These shifts occurred not only because of the scientific challenges of this iconoclastic newcomer, but also possibly because experiences (or lessons) suggested that such a new type of material might have unpredicted properties and applications, as nanoscale materials have shown in recent years for example. So far, some QCs and ACs appear to have potential merit as thermoelectrics [6], catalysts [7], hydrogen storage materials [8], photonic crystals [9], and bio-inspired materials [10], in addition to their traditional applications as specific alloys and surface coating materials [5]. However, it should also be noted that principally metallurgists, physicists, and material scientists have been involved in the development of most QC/AC systems, whereas relatively fewer chemists have been engaged in this emerging research field.

For a long period, all crystalline solids were believed to possess 3D translational periodicity, and their structures could be unambiguously solved with the aid of modern crystallographic tools. However, this situation was challenged by the discovery of QCs. After several years of debate as to whether QCs were twins of cubic crystals [11, 12] and of discoveries of other novel QCs with enhanced qualities, the International Union of Crystallography (IUCr) redefined *crystal* as "*any solid having an essentially discrete diffraction diagram*," and an *aperiodic crystal* as "*any crystal in which 3D lattice periodicity can be considered to be absent*." As a consequence, QCs fall into the category of aperiodic crystals [13].

Unlike crystals that are packed with identical unit cells in 3D space, aperiodic crystals lack such units. So far, aperiodic crystals include not only quasiperiodic crystals, but also crystals in which incommensurable modulations or intergrowth structures (or composites) occur [14]. That is to say, quasiperiodicity is only one of the aperiodicities. So what is quasiperiodicity? Simply speaking, a structure is classified to be quasiperiodic if it is aperiodic and exhibits self-similarity upon inflation and deflation by tau ($\tau = 1.618$, the golden mean). By this, one recognizes the fact that objects with perfect fivefold symmetry can exist in the 3D space; however, no 3D space groups are available to build or to interpret such structures.

To interpret QC structures, higher-dimensional crystallography is necessary [2, 14]. However, this is extremely abstract and inconvenient, and Fourier transformations in higher dimensions are still "black holes" to most scientists. Moreover, a resistance to a paradigm shift from the conventional 3D into higher-dimensional crystallography is inherent (even for Linus Pauling [11, 12]). Given here is only a general introduction to higher dimensions and their conversions to three dimensions, as exemplified by the projection of a 2D square lattice onto a 1D space. As shown in Fig. 1, the grid points (atoms) of the square lattice can be projected onto any 1D physical (or parallel) subspace R_{par} at an angle α with respect to the horizon rows of the square lattice. (R_{per} is the complementary or perpendicular space.) If the slope ($\tan \alpha$) of R_{par} is a rational value with respect to the square lattice, the projection on R_{par} is a 1D periodic structure. For example, if $\alpha = 45°$, R_{par} goes repeatedly through grid points of the square lattice, and the resulted projection in the R_{par} subspace gives equal separation for neighboring grid points. In contrast, if the slope is irrational, the projection on R_{par} no longer gives a periodic structure; the separation of neighboring grid points actually can be very small, meaning that a problem of unphysically short interatomic distances in the R_{par} subspace results.

One way to solve the problem of unphysically short atomic distances is to project onto the R_{par} subspace only those grid points included in a selected strip (gray area), with width of $a(\cos \alpha + \sin \alpha)$ in the R_{per} subspace. The slope of R_{Par}^2 shown in Fig. 1 is 0.618..., an irrational number related to the golden mean [$(\sqrt{5}+1)/2 = 1.618...$]. As a result, the projected 1D structure contains two segments (denoted as L and S), and their distribution follows a 1D quasiperiodic Fibonacci sequence [2] (c.f. Table 1). From another viewpoint, the 1D quasiperiodic structure on the R_{par} subspace can be conversely decomposed into periodic components (square lattice) in a (higher) 2D space. The same strip/projection scheme holds for icosahedral QCs, which are truly 3D objects but apparently need a more complex and abstract 6D

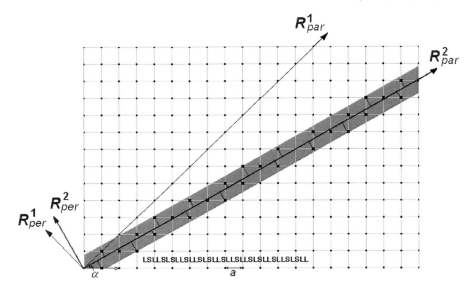

Fig. 1 Illustration of a two-dimensional square lattice (*filled circle*) projected onto one-dimensional space (R_{par})

Table 1 The Fibonacci sequence and its relationship to the golden mean, 1.618

Cycle	Sequence	L/S	Ratio
1	L	1/0	/
2	LS	1/1	1
3	LSL	2/1	2
4	LSLLS	3/2	1.5
5	LSLLSLSL	5/3	1.667
6	LSLLSLSLLSLLS	8/5	1.6
7	LSLLSLSLLSLLSLSLLSLSL	13/8	1.625
8	LSLLSLSLLSLLSLSLLSLSLLSLLSLSLLSLLS	21/13	1.615

The Fibonacci sequence can be generated by transformations of L→LS and S→L in each cycle. L/S represents the sequence of ACs that can exist for any QC system. With increasing order, the L/S ratio converges to the golden mean value

space to be fully described. Fortunately, the details of a QC probably do not convey too much in terms of structure and bonding; in one sense it is merely another new phase with an elemental composition and building blocks that are probably similar to those in the corresponding ACs. A glimpse of the roles of ACs vs a QC can be obtained from the relationship between their lattice parameters. According to high-dimensional crystallography, the cubic cell parameter ($a_{q/p}$) of a q/p AC is related to the QC lattice constant (a_6) by $a_{q/p} = 2a_6(p+q\tau)/(2+\tau)^{1/2}$, in which τ is the golden mean, and p and q are two consecutive Fibonacci numbers [0, 1, 1, 2, 3, 5, 8, 13, ..., $F(n) = F(n-2) + F(n-1)$] [15]. Accordingly, an i-QC can be considered as a cubic AC with an infinite lattice constant, and the higher the order (q/p) of an AC, the closer its structure approaches that of the i-QC. Therefore, ACs appear to be more important than QCs, at least in the beginning, because they offer knowledge about building blocks that are assumed to exist in the actual QCs. Thus, they play essential roles in modeling of QCs, whereas the QCs presumably will at best serve as justifications of the models when their structures are unambiguously solved through higher-dimensional crystallography.

To date, two decades after the first discovery of QC, there are still no more than 200 QC systems, and only one third of them contain thermodynamically stable QCs at room temperature [16]. Of all known QCs, i-QCs represent the largest group, the remaining being the 1- or 2D QCs. The last group includes octagonal (with eightfold rotational axes), decagonal (tenfold), or dodecagonal (12-fold) QCs. In the following, we will consider only QCs with quasiperiodicity in three dimensions, i.e., i-QCs.

So far, three types of i-QCs appear in the literature: Mackay [17], Bergman [18], and Tsai types [19], which have been differentiated on the basis of the polyhedral cluster sequences observed in the respective 1/1 AC structures. These are commonly represented as shown in Fig. 2. An i-QC is concluded to be Mackay-type if its 1/1 AC contains a 54-atom multiply endohedral cluster ordered, from the center out, as a small icosahedron (12 atoms), a larger icosahedron (12), and an icosidodecahedron (30). This motif occurs in ACs that consist of transition metals and main-group elements on the right side of the periodic table such as Al–(Pd,Mn)–Si [17, 20]. In

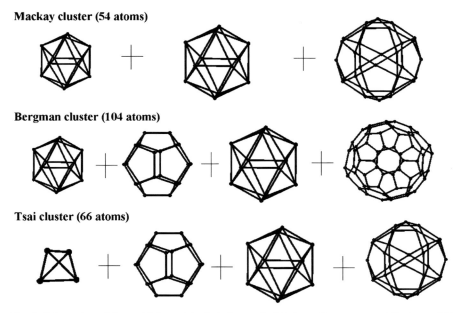

Fig. 2 Schematics of the multiply endohedral clusters in Mackay-, Bergman-, and Tsai-type 1/1 ACs

parallel, an i-QC belongs to the Bergman type if its 1/1 AC features four endohedral shells, from the center out, an icosahedron (12), a pentagonal dodecahedron (20), a larger icosahedron (12), and a cluster with a geometry similar to that of a Buckeyball (60). Usually Bergman-type phases are formed in so-called free-electron-like s, p systems, e.g., Al–Mg–Zn [18] and Li–Cu–Al [21]. Likewise, an i-QC is said to be a Tsai-type if its 1/1 AC contains the 66-atom cluster sequence of a disordered tetrahedron (4), a pentagonal dodecahedron (20), an icosahedron (12), and an outermost icosidodecahedron (30). Such a motif is commonly observed in structures with appreciable contributions of low-lying d states of the electropositive element, such as in (Yb,Ca)–Cd [19, 22, 23], in contrast to those in the Bergman types. However, not every 1/1 AC has the ideal cluster geometry; rather defect versions (fractional occupancies, split positions, interstitials etc.) are also encountered therein. For example, the innermost icosahedron of Bergman-type sometimes may be centered by a partial or full atom [18]. Additional interlayer atoms may also be found as interstitials between the penultimate and the outermost shells, e.g., in the Tsai-type clusters [24]. In this connection, the reader should be aware that strong bonding naturally also occurs between the neighboring shells. In addition, the second and third polyhedra in both Bergman and Tsai types, a dodecahedron and a larger icosahedron, are dual polyhedra and may alternatively be represented or assigned as a single intrabonded 32-atom triacontahedron. Finally, both Bergman and Tsai types evidently always occur within an additional larger triacontahedral shell. This aspect is considered further in Sect. 5.

Sometimes i-QCs are identified as P- (primitive) and F- (face-centered) types according to the symmetry of the i-QCs in 6D [25]. Assignments of these types are usually made on the basis of electron diffraction patterns, but one can evidently make an easier judgment according to the space group of the corresponding 1/1 AC. To date, 1/1 ACs of Mackay type always crystallize in space group $Pm\bar{3}$; different clusters at the origin and body center therefore are projected onto a face-centered superlattice in 6D space. In contrast, the same clusters at origin and body center for Bergman- and Tsai-type 1/1 ACs ($Im\bar{3}$) are projected onto a primitive lattice in 6D space. The interchange of lattice types between the 1/1 AC and the corresponding i-QC is noteworthy. According to higher-dimensional crystallography, I- (body-centered)-type i-QCs are also possible, but none has been reported so far. Sometimes i-QCs are grouped in terms of the major component, e.g., Al-, Cu-, Zn-, or Cd-based QCs [25]. This is just jargon and not very meaningful to chemists.

After the discovery of the Al_6Mn i-QC [1], development of QCs were limited for almost a decade to ternary systems with a major Al constituent, such as Al–(Pd,Mn)–Si, Al–Zn–(Li,Mg), Al–Cu–TM (TM = Fe, Ru, Os), Al–Pd–(Mn,Re) [2,25,26]. (This may be the reason why jargon such as "Al-based QCs" was coined.) After all, most QC discoveries were achieved by chemical additions to, or substitutions in, known compounds. From the mid-1990s to about 2000, QCs were also found in Zn–Mg–R (R = rare-earth-metal), Cd–Mg–R, and (Yb,Ca)–Cd systems, the last being the first stable binary i-QC at room temperature. Experience and insight are worth a lot — Tsai and coworkers produced $\sim$90% of these i-QCs [27].

The discovery of the binary (Yb,Ca)–Cd i-QCs [19] was a remarkable milestone in the history of QCs. The reasons are apparent: they offered unique opportunities for structural analyses as they exhibited negligible chemical disorder, probably because of the large differences in the chemical crystallography of the components, in contrast to more common problems with ternary intermetallics. In addition, they also represented new (Tsai) types of AC cores and of i-QCs with a structural motif different from those of the Mackay and Bergman types (above) that were better known at the time. Without doubt, such a breakthrough discovery must lead to an era of related chemical explorations or tunings. Actually, almost all of the i-QC systems developed since 2000 are Tsai types [28, 29], including our own additions (below).

Before we started synthetic explorations for novel i-QCs and ACs in 2001, there was general agreement that Hume–Rothery rules played essential roles in the stabilization of QCs [25]. Hume–Rothery rules [30, 31] were originally generalized from and used to guide research on novel sets of relatively electron-poor intermetallic compounds ($e/a < 2.0$) with some phase widths in which matching of atomic sizes, electronegativities, and valences between the component atoms appeared necessary. Later, these rules were further developed by Jones [32] to describe the electronic stabilization mechanism of "electron phases," that is, a size match between Fermi sphere ($2k_F$) and Brillouin zone (K_{hkl}). Indeed Hume–Rothery rules benefited Tsai and coworkers [25, 27] a lot in their systemic searches for QCs; they confined the e/a values and atomic size ratios to certain values. In theory, the mixing of low-lying d orbitals of electropositive elements with broader s and p bands of the more electronegative components near the Fermi energy (E_F) was recognized as an important factor in the stability of i-QC phases according to a study of the $(Ca,Yb)Cd_6$ 1/1 ACs [33]. Mizutani et al. [34] also illuminated the interplay between atomic and electronic structures from a theoretical study of various Bergman- and Mackay-type 1/1 ACs, as well as reasons why a Hume–Rothery mechanism works in QC systems. Their studies validated the significant contributions of s, p, and d orbital mixing around E_F to the stabilization of QCs, in addition to the Fermi sphere–Brillouin zone effects.

By 2001, all known QCs and ACs were intermetallic compounds, meaning that they are combinations of two or more metals, and metal clusters and metal–metal bonding dominate in their structures. (Recently, QCs have also been observed in supramolecular dendritic liquids [35].) However, about 75% elements in the periodic table are metals, so the possible combinations of metallic elements can afford a very large number of binary, ternary, and quaternary intermetallic compounds. Among these must be the very select groups and compositions that will generate quasicrystalline phenomena, although the latter are predictable only by extrapolation. However, questions as to which systems, which elemental proportions, and what reaction conditions will produce new QC/ACs were and are still open. Moreover, not many chemists were well-enough informed about QC/ACs to guide these quests, and so little chemistry has been pursued in this area.

2 Traditional Metal Clusters

2.1 Halide, Oxides, and Sulfide Systems

Historically, polyhedral clusters containing strong metal–metal bonding appear mainly in well-reduced compounds of transition metals with a few valence electrons in d (plus other) orbitals that lead to filled molecular orbital levels for the metal clusters. Such clusters occur frequently in transition metal halides, oxides, and sulfides, etc. The sizes of these clusters are conceptually limited by the space required by the counteranions that are bonded to the cluster surfaces, a common premise being that the clusters would condense further were unprotected metal atoms to contact each other. Some of the earliest examples include (i) $(M_6X_{12})^{n+}(X^-)_n$ for M = Nb, Ta, X = Cl, Br, I and $n = 2$, 3, 4, and (ii) $(Mo_6X_8^{4+})X^-)_4$ (and a few W analogs) [36, 37]. Another group of strongly bonded cluster phases are found in condensed, metallic, and interstitially stabilized rare-earth-metal halides in which one of many different transition metals are strongly bound within each host octahedron, e.g., $Pr_4(Ru)I_5$ [38]. There are also considerable varieties of condensed or intergrown Nb(Ta) cluster oxides and chalcogenides and their ternary metal derivatives [39], including some superconducting (i.e., Chevrel) phases Mo_6Ch_8 (Ch = S, Se) in which the intercluster metal–metal bonding is relatively weak [40].

Nevertheless, no AC or QC is known in any of these systems. After all, clusters in these halide, oxide, and sulfide systems consist of three to, most frequently, six metal atom aggregates, far from icosahedral symmetry. Many of the interstitial-free phases are evidently also semiconductors and therefore not good prospects for strong extended metal–metal bonding, although e/a values (counted over the metal atoms only) are as low as 2.3–2.5 in the most reduced phases. However, several key features are different in all. First, filled and low-lying anion bands are always present, that is, the electronegative atoms do not participate well in overall "intermetallic-like" bonding between all component atoms. The anions may also lead to "matrix effects" by which metal–metal distances are restrained if not increased [41]. Of course, there are also many 2D or 3D metallic compounds that lack easily recognizable clusters, such as the simple TiO and LaS (NaCl-type structures), but the anions in all of these also appear to be too electronically segregated to afford interesting QC/AC phenomena [42].

2.2 Zintl Phases

More often, polyhedral clusters with strong metal–metal (or metalloid–metalloid) bonding are the major structural motifs of classic Zintl phases. These are nominally salts composed of reduced p- (i.e., post-transition) elements that are usually interbonded into closed shell polyanions plus active metal cations, originally the alkali metals. The latter are considered to simply donate their valence electrons to the anion construction [43].

The variety of polyanionic clusters in Zintl phases is large, and these usually exhibit electronic regularities that customarily result from octet rule (closed shell) bonding, viz., Wade–Mingos or Zintl–Klemm concepts. For example, the salts Mg_2Si, CaGe, LiGe, KSn, and LiAl will, with some thought, be seen to be consistent with the formation of closed shell anion structures of, respectively, Si^{4-} monoanions, zigzag chains (or rings) of two-bonded Ge^{2-}, three-bonded puckered Ge^- layers (as a stuffed As-type structure), three-bonded Sn tetrahedra $[(K^+)_4(Sn_4^{4-})]$, and four-bonded Al^- in a cubic diamond lattice. Cations are accommodated between these units.

Some pertinent Zintl phases contain isolated, condensed, or extended anionic frameworks of larger and more symmetric polyhedra, e.g., icosahedra or their dual, pentagonal dodecahedra, as in K_3Ga_{13}, Li_2Ga_7, MGa_7 (M = Rb, Cs) [44], which are commonly accepted as major building blocks of QC/ACs. However, no QC phase has been reported in these systems. After all, these structures are still relatively open, and the bonding between the countercations and the polyanionic substructures that accommodate the cations are more ionic in character, and the overall e/a values for these are higher (>2.5) than found in QCs. Yet these phases provide some clear ideas about what should be considered in the development of QC/AC systems, that is, to increase the covalent bonding of the cations with the polyanions and to lower e/a values relative to classic Zintl phases. This idea naturally led to our interest in polar intermetallics, as discussed in Sect. 2.3.

2.3 Polar Intermetallics

Polar intermetallics are loosely referred to as electron-poorer relatives of Zintl phases in which the active metals do not contribute all of their valence electrons, rather they bond with the more electronegative components to some degree. The structures cannot be simply accounted for by octet rules because of substantial delocalized bonding among the atoms.

The components of polar intermetallics generally include an active metal from the group 1 or 2 or the rare-earth series plus, sometimes, a late-transition metal, and a metal from the p-block. Because of the presence of an electron-poorer late transition metal, polar intermetallics generally have lower e/a values (about 2.0–4.0) than classic Zintl phases (>4.0) [45]. Note these values are traditionally calculated over *only* electronegative atoms [45], in contrast to those of Hume–Rothery phases (<2.0) [45] and QC/ACs (2.0 ± 0.3) [25], for which electron counts are considered to be distributed over *all* atoms. The former two higher values are decreased to about 1.5–2.5 and >2.5, respectively, when counted over *all* atoms (but with omission of any d^{10} shells). For comparison purposes, Fig. 3 sketches the distribution of all these intermetallic phases according to e/a counted over all atoms, as we will use hereafter.

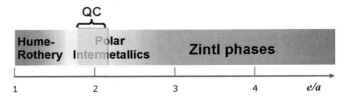

Fig. 3 Relative distributions of intermetallic phases in terms of e/a values

In particular, the structures of some polar intermetallics exhibit one of the requisite features of the AC/QC family (icosahedra and related symmetries) when p-elements utilized are moved right to left to the group 13 triels Ga, In, Tl [46]. For example, $Na_{14}K_6Tl_{18}M$ (M = Zn, Cd, Hg) contains $M@Tl_{12}$ icosahedral clusters [47], whereas $K_{34}Zn_{20}In_{85}$ [48] and $Li_{13}Cu_6Ga_{21}$ [49] contain the so-called Samson polyhedra that have been considered as important building blocks for QC modeling [50].

Since all known QC systems, with e/a of about 1.75–2.20 [25], lie close to the approximate border between the Hume–Rothery and polar intermetallic phase regions, a reasonable starting place for development of new QC/AC systems is to study selected polar intermetallic systems with nearby e/a values. Synthetic explorations of such polar intermetallics have been significant only in the past few decades [42, 45]. Knowledge and insights developed about the diverse interplays between composition–structure–electronic structure–physical properties for these phases were expected to be a considerable aid to the discovery of novel QC/ACs.

3 Toward Quasicrystals and Their Approximants

3.1 Polar Intermetallics Containing the Triels

Where might one find new AC/QC materials? Our earlier extensions of the p-element components in Zintl-type compounds to the triels (Al, Ga, In, Tl) (beyond the so-called "Zintl boundary") as well as the inclusions of late transition metals had already given us some significant glimpses of new and appropriate chemistry and structures [46, 51]. In addition, some examples were already in the literature for Ga [44]. In contrast, the anion or intermetallic chemistry of aluminum [45] stands apart from that of the three heavier group members (Ga, In, and Tl) and will not be considered here significantly.

The numerous cluster compounds found in A–Tr systems (A = alkali metal; Tr = Ga, In, Tl) progress from 3D networks with Ga to more isolated anionic clusters with Tl, with In being more or less intermediate to both [52]. These distributions also depend to some degree on the relative proportions of the A and Tr components. In any case, the considerable tendency for relatively strong and diverse homoatomic binding (catenation) among these triel elements in their negative oxidation states

A Chemical Approach to the Discovery of Quasicrystals

was quite unanticipated. Our principal investigations [46] have been with In, and secondarily with Tl, in part because there had already been substantial studies of A–Ga systems by Belin and coworkers [44], the results of which are largely distinctive from ours. Stable Zintl phase clusters of the tetrels (Si, Ge, Sn) with alkali metals commonly have charges of 2- or greater, evidently in order to provide a sufficient number of cations to prevent further condensation. Isoelectronic triel analogs of these would require an increase of n in the polyanion charge for n Tr atoms, and such high charge states for finite polyanions are unknown. Rather, a quite new cluster and network chemistry develops. The first were electron-deficient deltahedra, analogs of well known polyborane species $B_nH_n^{2-}$ that can be well-described by Wade's rules [43,53], etc., but their e/a values were still high. More significant were new hypoelectronic cluster examples (i.e., with $2n$ and $2n-4$ electron counts versus $2n+2$ for closo-deltahedra by Wade et al.). Some of the more provocative species, as far as the present subject matter is concerned, were salts of the new Tl_9^{9-} (a presumed icosahedral fragment) [54], Tr_{11}^{7-} (a new polyhedron) [55], and Tl-centered icosahedral $Tl_{13}^{10,11-}$ [56] anions together with heterometal-centered examples of the last [47]. In addition, the literature already contained examples of interbonded Ga_{12}^{2-} icosahedra in Li_2Ga_7 [57] and of dodecahedra in AGa_3 (A = K, Ca) [58,59], to a degree that this family of elements is sometimes referred to as the icosogens [60], the icosahedron-formers. Many of these "salts" are poor metals as far as conduction characteristics, a property that originates with the highest-lying electrons, whereas the many cluster structures we see originate from bonding with the more tightly bound electrons [51]. Furthermore, the calculated COOP or COHP values (overlap populations) for pair-wise bonding in these phases are frequently optimized, which suggests that the structures often attain maximum bonding at their particular electron populations.

Looking a little deeper into the periodic table discloses some structures that contain novel and highly condensed polyhedra as alkali metal salts of the "diels" Zn and Cd, $NaZn_{11}$ [61] for example, but our attempts to tune triel phases by addition of these so-called diel elements alone were not very rewarding. (These still remain challenging opportunities.) On the other hand, numerous A–(Cu, Ag, Au)–Tr systems have subsequently been discovered to be very productive in terms of new network types and structures, but not of the types sought here [46,62,63]. Note that no AC/QC examples of the alkali metals are known except with Li, the other members evidently being too electropositive to participate well in the bonding and delocalization. Instead, the smallest alkaline-earth metal Ca, plus Mg as well as the smaller or later rare-earth metals, are evidently more suitable components for bonding within AC/QC networks.

In more distant explorations, the frequent occurrences of Mg, Zn, Ca etc. in known QC systems also led us to reexamine some of the older reports of unusual phase diagram or structural results for the Ca–Zn, Mg–Zn, Mg_5Tr_2 systems as well as those of Ca_3AlCd_{17}- or $NaZn_{13}$-type for Ae–Zn, Ae–Cd, and $Ae(Cu,Ag)_6Tr_7$ (Ae = alkaline-earth metal) [61], including some ternary extensions. We also remained attentive to the occurrence of structures with space groups that are subgroups of icosahedral symmetry (below). However, nothing very unusual or encouraging was found, at least with our more-or-less superficial checks. Such outcomes naturally depend on the intuition applied regarding compositions or third elements investigated. Some of the hints regarding the unusual phases noted above were published some decades earlier and may have originated with inadequate structural (powder pattern) characterizations, but considerable room for further study still appears to exist.

The discovery of the icosahedra-based phases $Na_3K_8Tl_{13}$, $Na_4A_6Tl_{13}$ (A = K, Rb, Cs) [56], and $Na_{14}K_6Tl_{18}M$ (M = Mg, Zn, Cd, Hg) [47] gave us good insights into how to proceed. The first two are hexagonal ($R\bar{3}m$) and cubic ($Im\bar{3}$) salts of Tl-centered Tl_{12}^{n-} icosahedra, respectively, but the second is unusually tightly packed, as shown in Fig. 4a. The smaller Na ions lie on the threefold diagonal axes of this cubic cell and bridge between icosahedral faces of adjoining Tl_{13} groups. As commonly found in such compounds, the Na:K proportions and positions are well fixed without site mixing, and the strong, efficient bonding in fact leaves each cluster with a well-localized electron hole in its molecular orbitals. A second strong

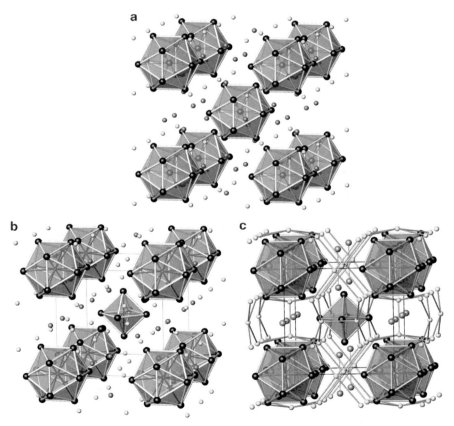

Fig. 4 Crystal structures of **a** $Na_4A_6Tl_{13}$ (A = K, Rb, Cs), **b** $Na_{14}K_6Tl_{18}M$ (M = Mg, Zn, Cd, Hg), and **c** Mg_2Zn_{11}, all of which contain centered icosahedral of the main group elements

point is that the space group is the same as later found in many 1/1 ACs, raising the question "how might the Na and K be replaced by cations that are more apt to participate in the desired intermetallic-like bonding in an AC and a QC?" A good clue came from a lucky attempts to substitute other centering elements for Tl, which gave $Na_{14}K_6Tl_{18}M$ (M = Mg, Zn, Cd, Hg), the primitive ($Pm\overline{3}$) cubic derivative with M-centered icosahedra and Tl_6^{8-} octahedra (Fig. 4b). (The change in Na:K proportions is exactly that necessary to put Na on all faces of the smaller cluster.) A strong hint from this last quaternary result came from the recognition that the group is isostructural with Mg_2Zn_{11} [$= (Mg_6)(Zn_{14})(Zn@Zn_{12})(Zn_6)$] (Fig. 4c), which is drawn for emphasis with networks of Zn_{14} that envelop the Zn_{12} and Zn_6 polyhedra and Mg_2 dimer spacers. This immediately suggested the greater covalency we are looking for in an intermetallic precursor to AC and QC. Tuning Mg_2Zn_{11} and other isotypic examples of this structure type with more suitable cations has been very successful (see Sect. 4). But, first we need to present important aspects of QC systems, properties, and symmetry in more depth.

3.2 Chemical Insights from Earlier Works

From Tsai's pioneering discoveries [25, 27], we know that atomic size, electronegativity, and valence electron counts play substantial roles in the formation of QCs. These criteria are expressed by the Hume–Rothery rules [30, 31]. However, three additional highlights are also important in the consideration of possible candidate systems, at least from the viewpoint of chemists.

Cluster types. According to 1/1 AC structures, all three i-QCs types contain building blocks with local fivefold symmetry (Fig. 2). Obviously, a search for possible new QC/AC systems should look to precursors in which clusters with pseudoicosahedral symmetries are major structural motifs.

To find some such structures is not difficult. Shoemaker and Shoemaker [50] have collected known binary intermetallic structures in which icosahedral clusters prevail, and they proposed some links between these phases and QC phenomena. In addition, studies of polar intermetallics and Zintl phases have revealed that a number of intermetallics containing heavier triels (Ga, In, and Tl) exhibit distinctive structural regularities [46]. For example, Tl in these often exhibit isolated empty or stuffed icosahedra, whereas Ga products often have extended frameworks of condensed icosahedra and related clusters, with In behaving in a more intermediate manner. A coloring of these apparent differences may still be present because of nonuniform investigations of possible systems, even as to the relative sizes and proportions of the usual alkali metal components employed. Moreover, some variations among icosahedral clusters may still occur when late transition metals (e.g., Cu, Ag, Au, Zn, Cd etc.) are incorporated. Therefore, polar intermetallics containing group 13 elements, the "icosogens," are in general good prospects as QC/AC candidates.

Active metal. The selection of active metal is also a critical factor. For polar intermetallics and Zintl phases, alkali, alkaline-earth, and rare-earth elements have been used, with parallel changes introduced by different (formal) charges on and proportions of the cations. So far, only Li among the alkali metals, Mg plus Ca in the alkaline-earth metals, and Sc, Y, and Yb have been shown to form i-QCs. The reasons must be manifold: size, d orbital contributions, valence electron counts, electronegativities, and bonding in specific structures, etc. Again, negative synthetic results are less apt to be reported.

A survey of literature reveals that active metals have essential influence on the types of QC/ACs formed. For example, Bergman-type QC/ACs to date contain only cations with free-electron-like s and p states around E_F (e.g., Mg and Li), whereas Tsai types are formed with active metals that have low-lying d states (Ca, Sc, Y, etc.).

Crystal Symmetry. Cahn etc. [64] has derived all the possible subgroups for icosahedral symmetry. According to group–subgroup relationships, the maximum crystallographic subgroup of the noncrystallographic icosahedral group ($m\overline{3}5$) is $m\overline{3}$, meaning that ACs and other precursors with $m\overline{3}$ symmetry are closer in symmetry (and hence structure) to i-QCs than those with other or lower symmetries. Therefore, one useful route to new QC/ACs is to check the crystal symmetries of diverse compounds with suitable elements to see whether they are appropriate. After all, the phase transformations from crystalline to QC phases must also obey Landau theory of phase transitions [65]. In practice, however, the most direct way to 1/1 ACs is to look in databases for possible precursors with $m\overline{3}$ point symmetry.

3.3 A Reward

Actually, before pseudogap tuning of certain good prospects (below) became a clear route for us, the tuning of $ScZn_6$ to QC/AC products guided by above considerations was useful. A check of the Pearson's handbook [61] led us to question the reported structure of "Sc_3Zn_{17}"($Im\overline{3}$) [66] because nothing was found within the innermost dodecahedral cage (~ 7.0 Å diameter); otherwise, the structure appeared to be isotypic with that of YCd_6 [67], the prototypic Tsai-type 1/1 phase [19]. In addition, Zn is a plausible icosogen, and Sc d orbitals are found to be important near E_F in theoretical studies of such intermetallic compounds. Therefore, we speculated that "Sc_3Zn_{17}" might be chemically tunable to a Tsai-type i-QC phase.

"Sc_3Zn_{17}" has e/a equal to 2.15, slightly larger than 2.0 for the (Ca,Yb)–Cd i-QCs, but substitution of some Zn by its neighbor Cu offers a ready opportunity to program e/a. Figure 5 shows the XRD patterns of reaction products with the composition $Sc_3Cu_xZn_{17-x}$ ($x = 0$–4), together with two simulations according to single-crystal refinements of Sc_3Zn_{17} and our Sc_3Zn_{18} ($= ScZn_6$) results [24]. Clearly, the reported structure of Sc_3Zn_{17} is wrong. Also, the patterns for $x < 3$ products are dominated by a new phase that was later identified as 1/1 AC, and that for $x = 4$ is a mixture of this 1/1 AC and the new i-QC phases. As more Cu was added (to lower e/a), the yields of the i-QC phase increased [24]. Single-crystal analysis also reveals that 1/1 AC is isostructural with YCd_6, not Sc_3Zn_{17} without the tetrahedron,

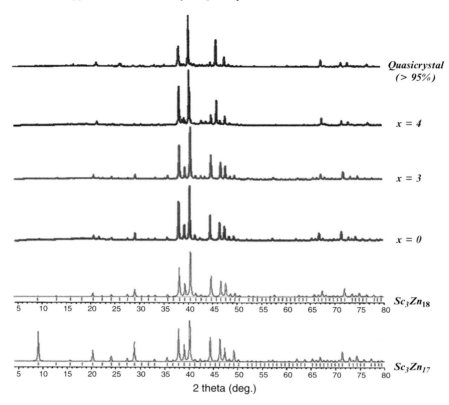

Fig. 5 XRD patterns for Sc$_3$Cu$_x$Zn$_{17-x}$ (x = 0–4), together with a pattern of pure i-QC (*top*) from a nominal Sc$_{15}$Cu$_{15}$Zn$_{70}$ reaction, and two patterns (*bottom*) simulated from the single-crystal data of the reported Sc$_3$Zn$_{17}$ and our Sc$_3$Zn$_{18}$. (Adapted with permission from [24]. Copyright 2004 American Chemical Society)

and has a composition of Sc$_3$Zn$_{\sim 18.0}$. But, in this system, Cu is evidently necessary for the formation of the i-QC, which has a composition of Sc$_{16.2}$Cu$_{12.3}$Zn$_{71.5}$ [68].

The successful tunings of Sc–Cu–Zn QC/AC demonstrated that our understanding and intuition about QC/ACs were in the right direction, which encouraged us to develop novel QC/AC systems from other intermetallic systems. We were in some sense lucky to have started these QC/AC tunings from Sc$_3$Zn$_{17}$. Such efforts not only supported our viewpoints on how to narrow the candidate precursors among intermetallics (according to cluster type, active metal, and crystal symmetry), but they also gave us direct experience in identifying a QC phase from powder pattern data and how to index its pattern with six integers (*not* three!). Although later experience showed us how to identify a QC more quickly (via observation of reciprocal lattice data sets collected on a STOE IPDS diffractometer or equivalent), it originally took about 3 months to confirm that the new phase we first obtained was an i-QC and not another phase. We first tried to index the QC powder pattern using

Elser's method [69] and also acquired Laue photos along the two-, three-, and five-fold axes rather than doing electron diffraction, which is usually beyond a chemist's laboratory.

Shortly after, we recognized that $ScCu_4Ga_2$ ($Im\bar{3}$) [70] might also be tuned to a QC, but the correct stoichiometry and reaction conditions were not achieved in our limited experiments. Recently, Honma and Ishimasa [71] have reported that i-QC phase forms almost exclusively from a rapidly quenched $Sc_{18}Cu_{48}Ga_{34}$ composition, emphasizing a very narrow phase width and its thermodynamic metastability at room temperature. However, the failure turned us to other Ga intermetallics, which led to the pseudogap tuning concepts that follow.

4 Pseudogap Tuning

4.1 Structure and Bonding in $Mg_2Cu_6Ga_5$

As noted above, it occurred to us that selected polar intermetallics containing triel elements, particularly Ga and In, under optimal conditions might conceivably be structurally, electronically, and chemically tunable to ACs and, eventually, to QCs. The promising Mg_2Zn_{11} family contained examples of triel Al in $Mg_2Cu_6Al_5$ [72], In in $Na_2Au_6In_5$ [73], and Tl in $Na_{15}K_6Tl_{18}M$ (M = Mg, Zn, Cd, Hg) [47], but nothing with Ga. A question naturally arose as to why no "$Mg_2Cu_6Ga_5$" had been reported in this family. Actually, a comparison of Mg–Cu–Al and Mg–Cu–Ga ternary phase diagrams revealed that although the two systems were quite similar, no Ga analogs of either $Mg_2Cu_6Al_5$ and $Mg_{32}(Cu,Al)_{49}$ phases had been found [74]. However, our synthetic explorations revealed that at least two more phases exist in the Mg–Cu–Ga ternary system; one is the sought $Mg_2Cu_6Ga_5$ [75], and the other is $Mg_{35}Cu_{24}Ga_{53}$ [76], a novel Laves-like phase that contains interpenetrating Bergman-like clusters.

Undoubtedly, $Mg_2Cu_6Ga_5$ exhibits same structural pattern as shown for the parent in Fig. 4c, that is, each unit cell contains three Mg dimers, a Cu_6 octahedron, a $Ga@Cu_{12}$ icosahedron, and a Ga_{14} network. From the viewpoint of QC structures, however, $Mg_2Cu_6Ga_5$ can alternatively be taken as primitive cubic packings of Cu_6 octahedra and 45-atom multiply endohedral clusters in an expanded cell, as shown in Fig. 6. Each 45-atom multiply endohedral cluster contains, successively from the center out, a Ga-centered Cu_{12} icosahedron, a $Ga_{12}Mg_8$ pentagonal dodecahedron, and a larger Ga_{12} icosahedron (with negligible intracluster Ga–Ga bonding). The geometry of this onion-like cluster is similar to that of the inner three shells of Bergman clusters (Fig. 2), which are arranged body-centered-cubic (b.c.c.) in $Mg_{32}(Al,Zn)_{49}$ [18] and R-Li_3CuAl_5 (both $Im\bar{3}$) [21]. Of particular interest, therefore, is the attainment of structures derived from or showing the higher $Im\bar{3}$ symmetry through substantial tunings. This is not just whimsical, however; a parallel structure change from the same primitive cubic packing ($Pm\bar{3}$, $x \leq 0.29$) to

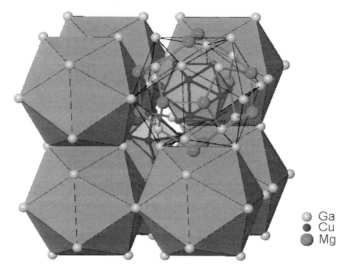

Fig. 6 The structure of $Mg_2Cu_6Ga_5$, emphasizing the primitive cubic packing of Cu_6 octahedra and three-shell 45-atom clusters. One of the latter is cut open to show its inner shells

b.c.c. packing ($Im\bar{3}$, $0.32 \leq\leq 0.69$) has been reported in $(Al, Si)_{82}Mn_xFe_{18-x}$ [77], the 1/1 AC of $Al_{74}Mn_{17.6}Fe_{2.4}Si_6$ i-QC [78]. The same occurs in reverse between $Na_4K_6Tl_{13}$ ($Im\bar{3}$) [56] and primitive cubic packing of comparable $Tl_{12}M$ units and Tl_6 octahedra in $Na_{14}K_6Tl_{18}M$ (M = Mg, Zn, Cd, Hg) [47], Fig. 4c.

A speculation led us to examine the electronic structure of $Mg_2Cu_6Ga_5$, by which we discovered a pseudogap in the DOS at $\sim$79 valence electrons (omitting Cu d^{10} electrons) with $e/a = 2.03$ (Fig. 7), just above E_F ($e/a = 1.92$) [75]. Moreover, the principal Cu–Cu, Cu–Ga, Ga–Ga interactions as described by COOP data ($\approx$overlap populations) are not fully optimized at E_F. Rather, Cu–Cu and Ga–Ga change bonding characteristics at energies slightly below the pseudogap, whereas other bonds would either require more electrons to be optimized or they remain bonding throughout according to this calculation.

Considering both the crystal and electronic structures, $Mg_2Cu_6Ga_5$ was therefore employed as a plausible precursor or a useful guide to a new QC/AC family; for which a conceptual pseudogap tuning under a rigid band assumptions suggests that four more electrons per cell in $Mg_2Cu_6Ga_5$ might push E_F into the pseudogap. It is to be noted that electronic tuning via compositional change is not new to solid-state chemists; however, those in the QC field have evidently not utilized this approach although an association of pseudogaps with QCs has long been established. Of course, we are also predicting a stable electronic composition across a presumed phase change, at least from $Pm\bar{3}$ to $Im\bar{3}$ in the 1/1 AC.

In Sect. 4.2 we expand on pseudogap tuning concepts and illustrate these ideas and applications to the isotypic $Mg_2Cu_6Ga_5$, Mg_2Zn_{11}, and $Na_2Au_6In_5$. Because all the ACs we have obtained have very similar structural motifs, their structural regularities will be discussed together later in Sect. 5.

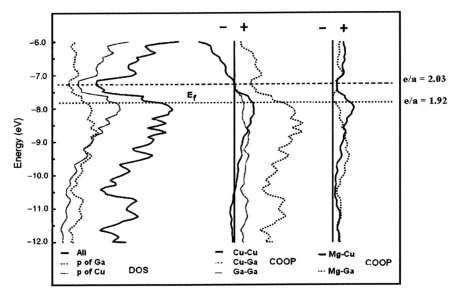

Fig. 7 DOS and COOP data for Mg$_2$Cu$_6$Ga$_5$ (EHTB). Note the pseudogap about 4 e/cell above EF. (Reproduced with permission from [75]. Copyright 2003 American Chemical Society)

4.2 Tuning ACs and QCs from Mg$_2$Zn$_{11}$-Type Precursors

4.2.1 From Mg$_2$Cu$_6$Ga$_5$ to Sc–Mg–Cu–Ga ACs and QC

Two approaches were considered to introduce four electrons per cell into Mg$_2$Cu$_6$Ga$_5$: (i) substitutions of electron-richer elements and (ii) variations of the elemental proportions. The latter approach was not successful during the order of 100 exploratory reactions run under diverse conditions. For instance, reactions of the modified Mg$_2$Cu$_{5.33}$Ga$_{5.67}$ with $e/a \approx 2.03$ yielded about 60% Mg$_2$Cu$_6$Ga$_5$, 30% MgCuGa, and 10% of the new Mg$_{35}$Cu$_{24}$Ga$_{53}$ [76]. On the other hand, pseudogap tuning of Mg$_2$Cu$_6$Ga$_5$ worked well when most of the Mg was replaced by Sc [79]. The selection of Sc arose not only because of its similar size and electronegativity to Mg and an additional valence electron, but also because of its low-lying d orbitals. Exploratory reactions of Sc$_{x/3}$Mg$_{2-x/3}$Cu$_6$Ga$_5$ within a limited e/a range (2.00–2.05) afforded very informative results, as revealed in part by the XRD patterns in Fig. 8.

Single crystals from the $x = 5$ reaction were found to crystallize in space group $Im\bar{3}$, $a = 13.5005(4)$ Å, with a refined composition of Sc$_3$Mg$_{0.17(4)}$Cu$_{10.5}$Ga$_{7.25(4)}$ or, normalized, as Sc$_{14.2}$Mg$_{0.8}$Cu$_{49.7}$Ga$_{34.3}$ [79]. This phase is isostructural with the 1/1 ACs (Ca,Yb)Cd$_6$ [22], meaning that an i-QC phase should exist in the neighborhood. Therefore, small variations in first Sc:Mg and then in Cu:Ga proportions, together with different reaction conditions, were tried and found to be fruitful. A quenched sample of Sc$_{15}$Mg$_5$Cu$_{47}$Ga$_{33}$ ($e/a = 2.01$) with a ~5% smaller

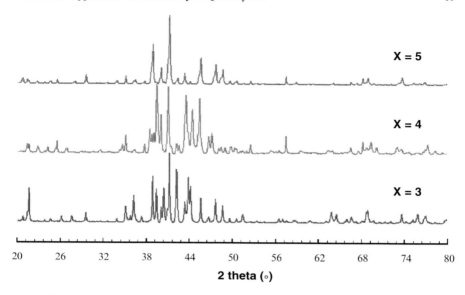

Fig. 8 XRD patterns for annealed $Sc_{x/3}Mg_{2-x/3}Cu_6Ga_5$ ($x = 3, 4, 5$) samples. The $x = 5$ sample is dominated by the 1/1 AC, $x = 4$, by a new incommensurately modulated 1/0–2/1–1/0 AC, and $x = 3$, by the 1/1 AC and a $Sc_{13}Zn_{58}$-type phase

(Cu+Ga) proportion consisted mainly of the QC phase, suggesting that this was the right direction for fine tuning. Accordingly, samples of $Sc_{15}Mg_3Cu_yGa_{82-y}$ ($y = 46$–48.5) were reacted. The i-QC was found exclusively for $y = 48$, as shown in Fig. 9, in agreement with the loaded QC composition reported by Ishimasa and coworkers [80]. Such a narrow composition range for this QC phase emphasizes that it could have been easily missed and, in the worst cases, false judgments could be made on the basis of too few "hunting" reactions (a common practice for solid-state chemists!).

As a check on the pseudogap predictions made for such a phase from $Mg_2Cu_6Ga_5$, the DOS and COHP for the hypothetical disorder-free "$Sc_3Cu_{10.5}Ga_8$" 1/1 AC model ($e/a = 2.02$ versus the refined value of 2.00) were calculated (LMTO). As shown in Fig. 10, a pseudogap remains at the same place in the DOS, with E_F located on a steep shoulder of the pseudogap, and bonding in the 1/1 AC is now optimized, at least as calculated for the Ga–Ga bonding. All of these are consistent with the predictions made earlier on the basis of DOS and COOP data for $Mg_2Cu_6Ga_5$ (Fig. 7). In addition, further examination of the bands for the 1/1 AC reveal that the pseudogap arises largely from mixing between s, p orbitals of Cu and Ga with the low-lying d of Sc and that these also enhance the depth of the pseudogap. All of these parallel earlier theoretical results for $(Ca,Yb)Cd_6$ reported by Ishii and Fujiwara [33].

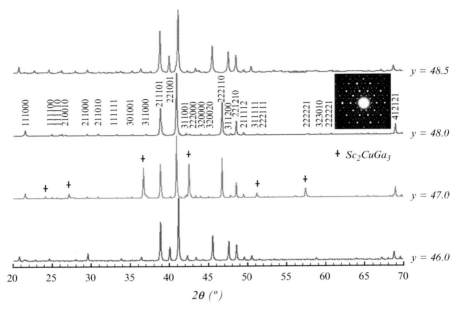

Fig. 9 XRD of quenched samples of Sc$_{15}$Mg$_3$Cu$_y$Ga$_{82-y}$. The $y = 46.0$ and 48.5 products are the 1/1 AC, that for $y = 48.0$ is the i-QC, and $y = 47.0$ results in a mixture of the QC and Sc$_2$CuGa$_3$. The *inset* and the six *integers* show the ED pattern of the i-QC along a fivefold axis and the six-dimensional indices. (Reproduced with permission from [79]. Copyright 2005 American Chemical Society)

4.2.2 From Mg$_2$Zn$_{11}$ to Sc–Mg–Zn ACs and QC

The tuning success with Mg$_2$Cu$_6$Ga$_5$ naturally encouraged a similar investigation of the parent Mg$_2$Zn$_{11}$ [81]. EHTB calculations for it revealed that empty bonding states remained for the major Zn–Zn bonding up to e/a of ~ 2.2, despite the fact that E_F already lay in a pseudogap [75] (Fig. 11). In order to fill the empty bonding states, exploratory reactions of Sc$_x$Mg$_{2-x}$Zn$_{11}$ over $e/a = 2.0$–2.2, were undertaken. Surprisingly, we obtained not only the thermodynamically stable 1/1 AC and i-QC phases, but also the valuable higher order 2/1 AC [82]. Figure 12 shows a portion of the Sc–Mg–Zn phase distributions according to X-ray powder data analyses. As can be seen, the Sc–Mg–Zn i-QC appears in this section in a dumbbell-shaped region, beyond which the 1/1 and 2/1 ACs exist. This particular i-QC forms over a wide composition range, in contrast to that in Sc–Mg–Cu–Ga [80]. We currently have little control over such subtle differences in electronic structure and packing that are presumably responsible for this.

As before, the 1/1 AC [83] crystallizes in space group $Im\bar{3}$ and is isostructural with YCd$_6$ [67], whereas the 2/1 AC [83], $Pa\bar{3}$, is isostructural with (Ca, Yb)$_{13}$Cd$_{76}$ [23]. All of these crystals have compositions similar to that of the i-QC. The normalized compositions for 1/1 AC ($x = 1.82$), 2/1 AC ($x = 0.75$), and i-QC

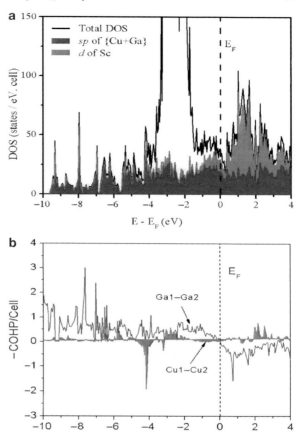

Fig. 10 DOS **a** and COHP **b** data for a hypothetical "Sc$_3$Cu$_{10.5}$Ga$_8$" 1/1 AC model ($e/a = 2.02$). (Reproduced with permission from [79]. Copyright 2005 American Chemical Society)

($x = 1.5$) are Sc$_{14.4}$Mg$_{0.9}$Zn$_{84.8}$ ($e/a = 2.15$), Sc$_{12.8}$Mg$_{2.9}$Zn$_{84.3}$ ($e/a = 2.13$), and Sc$_{14.6}$Mg$_{3.3}$Zn$_{82.1}$ ($e/a = 2.15$), respectively.

LMTO calculations on a hypothetical ScZn$_6$ 1/1 AC [82] revealed that Sc plays the same role here as in Sc–Mg–Cu–Ga 1/1 AC (Fig. 13). The Sc not only provides valence electrons to push E_F into the pseudogap, but its d orbitals also afford mixing with Zn s, p orbitals to enhance the depth of the pseudogap. This may explain why no Mg–Zn binary or Mg–Cu–Ga ternary Tsai-type QCs exist, but the Sc–Cu–Zn i-QC [24,68] forms, although its discovery was not directed by the pseudogap tuning concept.

4.2.3 From Na$_2$Au$_6$In$_5$ to (Ca,Yb)–Au–In ACs and QCs

As before, the DOS and COOP functions calculated for Na$_2$Au$_6$In$_5$ [84] exhibit a pseudogap and empty Au–In and In–In bonding states above E_F (Fig. 14). The

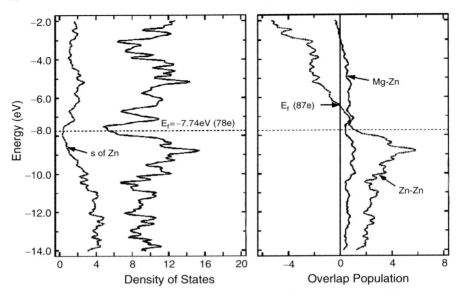

Fig. 11 DOS and COOP data for Mg_2Zn_{11}. (Reproduced with permission from [75]. Copyright 2003 American Chemical Society)

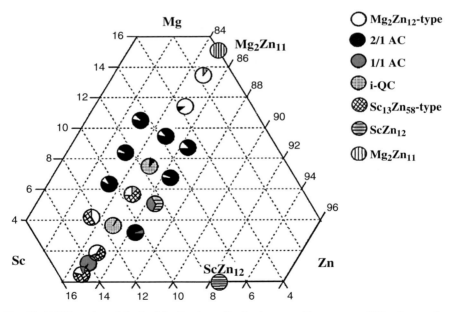

Fig. 12 400°C section of the Sc–Mg–Zn phase distribution according to power diffraction results, roughly outlining the phase regions of the 1/1 AC, 2/1 AC and Tsai-type i-QC

A Chemical Approach to the Discovery of Quasicrystals   23

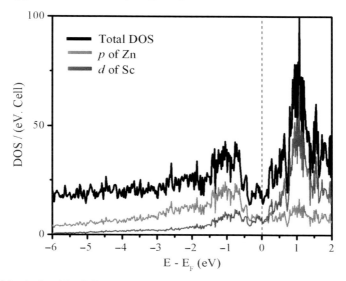

Fig. 13 DOS calculated for ScZn$_6$

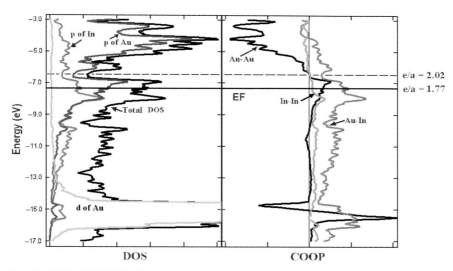

Fig. 14 DOS and COOP for Na$_2$Au$_6$In$_5$. (Reproduced with permission from [79]. Copyright 2005 American Chemical Society)

replacement of Na by electron-richer active metals that are more appropriate to QC bonding (e.g., Mg, Ca, Sc, or rare-earth metals) was considered in order to push the E_F ($e/a = 1.77$) into the pseudogap ($e/a = 2.02$). According to earlier works, however, there is strong theoretical evidence that d orbitals are critical for Tsai-type QCs, and atomic sizes also seem important. Therefore, Mg and Sc were taken out of consideration for these respective reasons. We first tried Ca and then Yb, encouraged by the idea that the Cd functions in (Ca,Yb)–Cd QC [19] systems might be replaced by mixed Au/In networks.

Luckily, the first exploratory synthetic reaction "$Ca_2Au_6In_5$" yielded primarily a pattern similar to that of the Sc–Mg–Zn 2/1 AC. This event suggested that the corresponding 1/1 AC and i-QC should also exist in this system, and that the supposed 1/1 AC should crystallize in the YCd_6-type structure. Further explorations via reactions of $CaAu_xIn_{6-x}$ were run to cover e/a over 1.6–2.3. Phase analyses revealed that these were very fruitful, affording single phase samples of 1/1, 2/1 AC, and the i-QC (Fig. 15). In addition, we also obtained $Ca_4Au_{10}In_3$, a novel phase with an unusually low e/a of 1.59, and wavy gold sheets [85].

As expected, the 1/1 and 2/1 ACs crystallize in b.c.c. and primitive cubic space groups $Im\bar{3}$ and $Pm\bar{3}$, respectively. The refined compositions for the 1/1 is $Ca_3Au_{12.2(1)}In_{6.3(2)}$ or, normalized, $Ca_{14.0}Au_{56.7(5)}In_{29.3(8)}$ ($e/a = 1.73$), and for the 2/1, $Ca_{12.6(1)}Au_{37.0(2)}In_{39.6(6)}$ or, normalized, $Ca_{14.1(1)}Au_{41.5(2)}In_{44.4(6)}$ ($e/a = 2.03$) [84]. It is noteworthy that e/a of the i-QC (1.98 by EDX) is very close to that predicted from calculations on $Na_2Au_6In_5$ (2.02) and the 1/1 AC (2.00); the DOS for the latter is shown in Fig. 16. It is well-known that Ca and Yb are interchangeable in many solid state systems because they have similar radii and the same common valence electron count in reduced systems. This trend continues in these systems; thus analogous Yb products were obtained from reactions similar to those used for the Ca–Au–In system.

4.2.4 From Ca–Au–In to Ca–Au–Ga Systems

The idea of searching QC in the Ca–Au–Ga system originated not just from the considerable chemical similarities of the two triels, but also from structural problems with the Ca–Au–In 1/1 and 2/1 ACs above, in which substantial amounts of configurational (positional) disorder and fractional occupancies occurred. Recall also that such disorder in the Sc–Mg–Zn ACs [83] is minimal. Moreover, these defects are thought to have little to do with the sample history, but possibly result from the smaller differentiation between the gold and indium both radially (size) and chemically. Therefore, replacement of indium by smaller gallium might provide a superb playground in which the effects of size on structures could be studied under some degree of isolation. Moreover, gallium and indium in many cases have similar electronic contributions in polar intermetallics, as they have same valence electron counts and a small difference in Mulliken electronegativities (Ga: 3.2 vs In: 3.1 eV) [86].

Earlier experience has shown that stoichiometric reactions of AB_xC_{6-x} (A: electropositive metal, B and C: electronegative main-group or late transition metals) with e/a of about 1.7–2.3 are a good rule-of-thumb way to generate Tsai-type AC/QCs. This time, suitable reactions of $CaAu_xGa_{6-x}$ provided both 1/1 and 2/1 ACs [87]. In addition, investigations of phase widths for both ACs also led to the discovery of the first Al-free 1/0 AC, $CaAu_3Ga$ [88].

A Chemical Approach to the Discovery of Quasicrystals

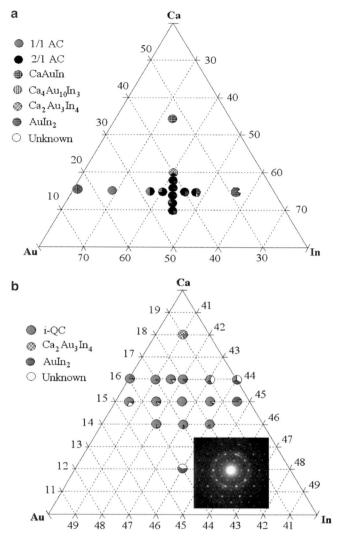

Fig. 15 Portions of the Ca–Au–In phase diagram, showing phase distributions for **a**: samples slowly cooled to 500°C and equilibrated, and **b**: samples quenched from 800°C over a more limited composition region. (Reproduced with permission from [79]. Copyright 2005 American Chemical Society)

Figure 17 shows a portion of the composition–phase distribution obtained from Ca–Au–Ga samples that were either slowly cooled from melts or annealed at 500°C for 3 weeks and quenched. Both ACs have considerable phase widths in Au/Ga proportions relative to previous studies, as reflected in the variations of lattice parameters [14.6732(7)–14.790(1) Å for the 1/1 AC and 23.8829(9)–23.9816(9) Å for the 2/1

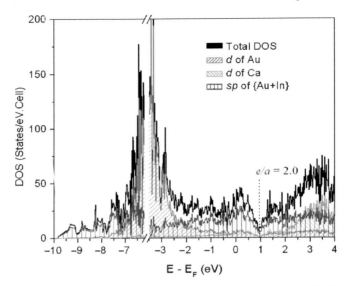

Fig. 16 DOS calculated by LMTO for the known Ca–Au–In 1/1 AC ($e/a = 1.73$). The pseudogap shown at $e/a = 2.00$ is close to the composition of i-QC phase, with $e/a = 1.98$. (Reproduced with permission from [79]. Copyright 2005 American Chemical Society)

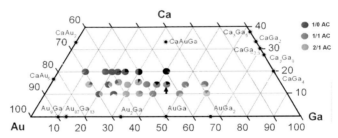

Fig. 17 Section of Ca–Au–Ga phase diagram showing the distribution of 1/0, 1/1, 2/1 ACs in samples that were slowly cooled from 800°C and annealed at 500°C. Note that the i-QC is obtained only in quenched samples near CaAu$_3$Ga$_3$ (marked by *the* arrow)

AC]. Structure determinations revealed that the 1/1 AC, Ca$_3$Au$_{\sim 9.3-12.1}$Ga$_{\sim 9.7-6.9}$, exists over about 42–55% of Au, whereas the 2/1 AC Ca$_{13}$Au$_{57.1(2)}$Ga$_{23.4(4)}$ may contain up to about 61% of Au. This means that the 2/1 AC has a smaller e/a value than the 1/1 AC in the Ca–Au–Ga system, a relationship opposite to that for the Ca–Au–In ACs. Notice also that the 1/1 AC for the first time has a Ca:(Au + Ga) proportion close to 3:19 rather than 3:18 because interlayer interstitial sites [Wyckoff 8c ($^1/_4$ $^1/_4$ $^1/_4$)] now have full occupancies. The reason is mainly related to the size contrast between Ga and In, the effect of which will be discussed in more detail in Sect. 5.

A Chemical Approach to the Discovery of Quasicrystals

5 Structure Regularities

Historically, the structures of ACs play critical roles in QC modeling: they provide important information on both the long- and short-range orders, i.e., the linkages among the outmost clusters and the atomic order within them. However, various views about clusters and definitions can sometimes raise enough confusion that conclusions such as "it is useless (or misleading) to give a description of the whole (QC) structure" have appeared as the summary of a panel discussion during the Ninth International Conference on Quasicrystals (ICQ9) [89]. The reasons may on one hand relate to the structural complexities of ACs (and perhaps also to the quality of the refinements achieved) and, on the other hand, to diverse structural descriptions that different authors have given for nominally the same AC structures. For example, Bergman-type structures have received at least three different structural descriptions. In the original report of Bergman et al. [18], the structure of $Mg_{32}(Al, Zn)_{49}$ was discussed in terms of four-shell 105-atom multiple clusters, but the same structure was described by Henley as a periodic packing of rhombohedral cells [90]. Likewise, the structure of the isotypic R-Li_3CuAl_5 [21] was given in terms of a larger b.c.c. packing of triacontahedral clusters, each of which contained the smaller Bergman-type collection of clusters.

To resolve the confusion, it is important to keep structural descriptions for all ACs and their corresponding QC models consistent. In other words, building blocks in ACs and corresponding QC models should be kept the same because the former are commonly assumed to contain the same or similar building blocks as the latter. According to our and other structure analyses [21–23, 79, 83, 84, 91], the larger triacontahedral clusters are the complete basic building blocks for both Bergman- and Tsai-type ACs rather than the smaller Tsai or Bergman polyclusters employed (below). In addition, a recent partial structural determination for the actual Yb–Cd i-QC [92] also reveals that triacontahedral clusters are indeed the major building blocks.

5.1 Tsai-Type ACs

All of these 1/1 ACs appear to have similar structural motifs and to be isostructural with YCd_6, with four-shell 66-atom Tsai-type clusters (Fig. 2) within b.c.c. triacontahedra. The detailed stoichiometries may still be slightly different because of structural defects or extra interstitial atoms. Similar Tsai clusters are also found in the 2/1 ACs (Fig. 18). However, such structural descriptions, although built of the customary cluster-based frameworks, may still contain many outer atoms that are not part of any Tsai cluster shell. These have been called "glue" atoms in a jargon of the QC field. The number of such glue atoms in a Tsai-type 1/1 AC cell is 36, about 21.4% of the total atoms, and this proportion increases in higher order 2/1 ACs (218 and 25.8%). In contrast, if the triacontahedra that encapsulate the original Tsai clusters are included as building blocks, then there are *no* glue atoms in 1/1 ACs, and only ~1.1% of the so-called glue atoms (interstitial dimer) remain in the

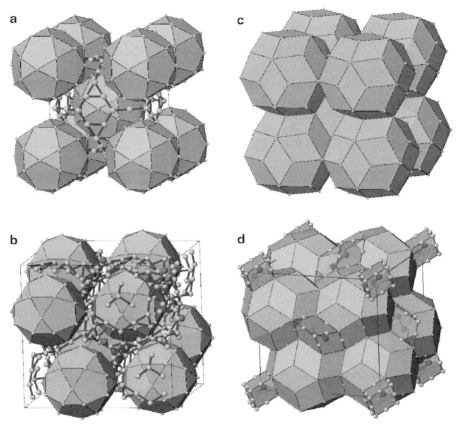

Fig. 18 Structural representations of Tsai-type **a** 1/1 AC, and **b** 2/1 AC plus the so-called glue atoms beyond the customary four-shell building blocks. In comparison, no glue atoms remain in **c** the 1/1 AC built from stuffed triacontahedra, and only dimers remain in **d** the 2/1

2/1 ACs. Moreover, the glue atoms beyond triacontahedra, in fact, center prolate rhombohedron (a cube elongated along a threefold axis), the recognized interstitial cavity among primitive cubic packed triacontahedra (also shown in Fig. 18).

Because a regular triacontahedron can be geometrically decomposed into ten prolate and ten oblate rhombohedra, *the 1/1 and 2/1 ACs can also be viewed as two different types of periodic condensations of prolate and oblate rhombohedral building blocks*. In this way, a link between AC structures and 3D Penrose tiles [93] used for i-QC modeling becomes evident. Therefore, the local atomic orders within and the linkages among triacontahedra are very useful in QC modeling.

Let's first consider local ordering of atoms within triacontahedra of Tsai-type ACs. The geometries for inner four shells are shown in Fig. 2, whereas the outermost triacontahedral shells appear in Figs. 19 and 20. Table 2 lists the chemical contents

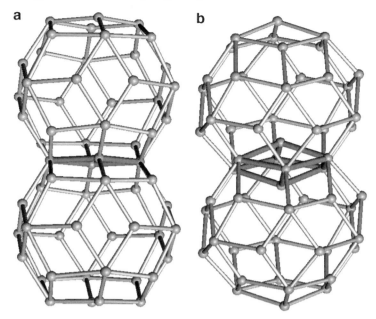

Fig. 19 The linkages between neighboring triacontahedral clusters in the Tsai-type 1/1 ACs along **a** the twofold axes, and **b** the threefold axes. The 2/1 ACs exhibit same linkages although they have different atom identities and symmetries. All decoration atoms of the triacontahedra are omitted for clarity. (Adapted with permission from [83]. Copyright 2006 American Chemical Society)

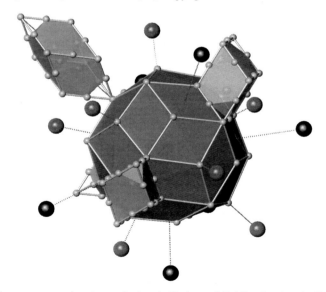

Fig. 20 Environment around a triacontahedron in Tsai-type 2/1 ACs, showing the linkages among triacontahedra and prolate rhombohedra. For clarity, only the central triacontahedron is shown as a polyhedron; the other 13 like neighbors are represented by smaller spheres at their centers. The same scheme holds for Figs. 21–23. (Reproduced with permission from [83]. Copyright 2006 American Chemical Society)

Table 2 Comparison of shell and PR contents within Tsai-type 1/1 and 2/1 ACs for different systems

System	Sc–Mg–Zn		Ca–Au–In		Ca–Au–Ga	
ACs	1/1	2/1	1/1	2/1	1/1	2/1
Formula	$Sc_3Mg_{0.2}Zn_{17.7}$	$Sc_{11.2}Mg_{2.5}Zn_{73.6}$	$Ca_3Au_{12.2}In_{6.3}$	$Ca_{12.6}Au_{37.0}In_{39.6}$	$Ca_3Au_{12.1}Ga_{6.9}$	$Ca_{13}Au_{57.1}Ga_{23.4}$
e/a	2.14	2.13	1.74	2.01	1.76	(1.64)
1. Tetrahedron	Zn_4	$Zn_{2.9}$	In_4	$In_{3.5}$	$Au_{2.8}Ga_{1.2}$	$Au_{1.2}Ga_3$
2. Dodecahedron	Zn_{20}	$Zn_{19.4}$	$Au_{15.3}In_{4.7}$	$Au_{10.2}In_{9.3}$	$Au_{11.3}Ga_{8.7}$	$Au_{13.3}Ga_{6.7}$
3. Icosahedron[a]	**Sc_{12}**	**$Sc_{9.6}Mg_{2.4}$**	**Ca_{12}**	**Ca_{12}**	**Ca_{12}**	**Ca_{12}**
4. Icosidodecahedron	$Zn_{29.2}Mg_{0.8}$	$Zn_{29.3}Mg_{0.7}$	$Au_{20.7}In_{9.3}[In_{3.8}]^b$	$Au_{11.0}In_{19.0}[In_{1.6}]^b$	$Au_{23.9}Ga_{6.1}[Ga_8]^b$	$Au_{26.1}Ga_{3.9}[Ga_9]^b$
5. Triacontahedron	Zn_{32}	Zn_{32}	$Au_{6.8}In_{25.2}$	$In_{29.6}$	Au_8Ga_{24}	$Au_7Ga_{21.9}$
	$(Zn_{60})^c$	$(Zn_{60})^c$	$(Au_{54.0}In_{6.0})^c$	$(Au_{48.0}In_{12.0})^c$	$(Au_{56.9}Ga_{3.1})^c$	$(Au_{57.3}Ga_{2.7})^c$
PR content		Sc_2		$Ca_{1.3}$		Ca_2
Reference	[83]	[83]	[84]	[84]	[87]	[87]

[a]Dominant cation sites are in bold type
[b]Interstitials in cubes between icosidodecahedral and triacontahedral shells are given in square brackets
[c]Additional decorations at or near the center of the triacontahedral edges are shown in parentheses

A Chemical Approach to the Discovery of Quasicrystals

of each shell in Tsai-type 1/1 and 2/1 ACs in the Sc–Mg–Zn [83], Ca–Au–In [84], and Ca–Au–Ga [87] systems. Several features among these details are remarkable.

1. In the inner four shells, only vertices are occupied by atoms, but for the outmost triacontahedral shells, not only the 32 vertices, but also the midpoints of (or close to) the 60 edges are occupied.
2. All the active metals in Tsai-type clusters lie on and define only the icosahedral shell. As a result, the global percentages of the active metals in Tsai-type phases are always $\sim 15\%$, in contrast to that in Bergman-type phases ($\sim 32\%$, below).
3. The innermost so-called tetrahedral shell in the 1/1 ACs, generated from a 24g position with about 1/3 occupancy, always refines as four atoms, but fewer in the 2/1 ACs. In contrast, tetrahedra in the 2/1 ACs are not centered in the dodecahedral cavities but are shifted along the unique threefold axis. In response, the neighboring dodecahedral shell exhibits strong distortion along that threefold axis. In addition, short distances between vertexes of the tetrahedron and the neighboring dodecahedron remain and these evidently require partial occupation of some vertex sites. This is the reason why the dodecahedral shells of 2/1 ACs generally refine with fewer than 20 atoms.
4. With an increase in size of the active metals, the interlayer interstitials between the triacontahedral and the penultimate icosidodecahedral shells appear to be occupied by smaller electronegative components, with variable occupancies. These interlayer interstitials are actually the centers of cubes and correspond to the Wyckoff 8c ($^1/_4$ $^1/_4$ $^1/_4$) special position in 1/1 ACs. Strictly speaking, occupation at this site means that the structure is no longer YCd_6-type but, for convenience, they are still referred to as Tsai-type phases. According to Piao and coworkers [94], occupation of these cube centers has strong correlation with the orientations of the innermost tetrahedra and distortions of the dodecahedra.
5. In the 2/1 ACs, the few glue atoms remaining are always dimers of the active metal in the PRs, and their occupancies may be either fractional or full. In the 1/1 ACs, each triacontahedron has $(8 + 6)$ neighbors, eight of which share ORs with the center one along on the threefold axes and the other six share rhombohedral faces along twofold axes (as distinguished in Fig. 19). The same linkages appear in 2/1 ACs, except that some lie along only pseudo two- or threefold axes. In contrast, each triacontahedron in a 2/1 AC has 13 like neighbors (Fig. 20). Of these, six locate on the pseudo twofold axes of the center triacontahedron and share rhombohedral faces with it; the other seven neighbors lie on the proper or pseudo threefold axes of the center cluster and share oblate rhombohedra with it. The decrease of like neighbors from 14 in 1/1 AC to 13 in 2/1 AC is compensated by the appearance of the interstitial prolate rhombohedra between the simple cubic packed triacontahedra. Four prolate rhombohedra surround each triacontahedron in the 2/1 ACs, as also shown in Fig. 20. Those on the proper threefold axes share vertexes with the center triacontahedron, and the other three on the pseudo threefold axes share rhombohedral faces. These arrangements are important long-range order (LRO) motifs.

5.2 Bergman-Type ACs

The uniform condensation of triacontahedra and interstitial prolate rhombohedra in Tsai-type 2/1 ACs offers some fresh points about LROs in i-QCs and stimulated us to consider whether the same linkages exist in Bergman-type 2/1 ACs. Bergman-type 1/1 ACs evidently also have triacontahedra as the outermost building blocks, as outlined in Table 3 for the Mg–Al–Zn system [91,95] and reported for the R-Al$_5$CuLi$_3$ 1/1 AC [21]. On the other hand, some structural data for the Bergman-type Mg–Al–Zn 2/1 AC in the literature are incomplete, either no positional data are given or the "glue" atom coordinates beyond the four-shell Bergman cluster are not reported [96,97].

Figure 21 shows the structure of Mg$_{27}$Al$_{11}$Zn$_{47}$ 2/1 AC [91], which was new inasmuch as this structure also exhibits the same triacontahedral and interstitial prolate rhombohedral clusters as building blocks as do the 2/1 Tsai types. Moreover, the LROs among these last two building blocks at the unit cell level and beyond are exactly the same as those in Tsai-type 2/1 ACs (Fig. 20), with only differences in short-range ordering within. Table 3 compares and contrasts shell contents, geometries, and sizes between a 1/1 and this 2/1 Bergman ACs vs the Sc–Mg–Zn 1/1 and 2/1 Tsai-type ACs [83]. The implied structural and bonding information therein is very impressive:

1. Bergman- and Tsai-type clusters have the same geometric types for the second, third, and fifth shells, with the innermost and the penultimate shells being different. Particularly, the third (icosahedral) shell and the outmost triacontahedral shell define comparably sized spheres for both, but the other three shells for Tsai types are about 1.0 Å smaller in diameter than for Bergman types.
2. The active metals have distinctively different locations in Bergman 1/1 and 2/1 ACs, Mg occupying the second dodecahedral and the outmost triacontahedral shells. These shells are occupied by more electronegative components in Tsai types, in which the active metals (Ca, Sc etc.) occur only in the third shell. Much chemical evidence can be found for the conclusion that Mg is the least acidic in the Mg–Zn–Al system (or forms the most basic oxide). Mulliken electronegativities have a very different basicity and are not suitable. These very fundamental differences (colorings) are of course intrinsic to the two structure types [45]. According to reported structural data, Tsai-type structures also form only in the presence of active metals that in principle contain low-lying d orbitals, whereas Bergman types form with cations lacking valence d orbitals. Theory is also in accord with these differences [33,98].
3. The dimers in the interstitial prolate rhombohedra have different electronic characters: the more electronegative elements in Bergman types, but active metals in Tsai types.

Table 3 Comparison of shell contents in the 1/1 and 2/1 ACs in Tsai-type Sc–Mg–Zn and in the Bergman-type Mg–Al–Zn structures

Shell	Polyhedron	Composition of Tsai-type ACs		Polyhedron	Composition of Bergman-type ACs	
		$Sc_3Mg_{0.2}Zn_{17.7}$ (1/1)	$Sc_{11.2}Mg_{2.5}Zn_{73.6}$ (2/1)		$Mg_{39.3}Al_{9.2}Zn_{51.5}$ (1/1)[a]	$Mg_{27.0}Al_{11.7}Zn_{47.3}$ (2/1)
1	Tetrahedron (1.50–2.2Å)[c]	Zn_4^b	$Zn_{2.9}^b$	Icosahedron (2.49–2.60 Å)[c]	Zn_{12}	Zn_{12}
2	Dodecahedron (3.40–4.05Å)	Zn_{20}	$Zn_{19.4}$	Dodecahedron (4.50–4.64 Å)	$\mathbf{Mg_{20}^d}$	$\mathbf{Mg_{20}^d}$
3	Icosahedron (4.89–4.93 Å)	$\mathbf{Sc_{12}^d}$	$\mathbf{Sc_{9.6}Mg_{2.4}^d}$	Icosahedron (5.0–5.20 Å)	$Zn_{9.2}Al_{2.8}$	Zn_9Al_3
4	Icosidodecahedron (5.60–5.78 Å)	$Zn_{29.2}Mg_{0.8}$	$Zn_{29.3}Mg_{0.7}$	Buckyball (6.63–7.29 Å)	$Mg_{12}Zn_{36.9}Al_{11.1}$	$Zn_{46.1}Al_{12.5}$
5	Triacontahedron (7.31–8.25 Å)	$Zn_{32}(Zn_{60}^e)$	$Zn_{32}(Zn_{60}^e)$	Triacontahedron (7.61–8.29 Å)	$\mathbf{Mg_{32}^d}$	$\mathbf{Mg_{32}^d}(Zn_{1.4}^e)$
PR content			Sc_2			Al_2
Reference		[83]	[83]		[95]	[91]

[a]Results for the 1/1 Bergman composition closest to that of the 2/1 AC

[b]Disordered tetrahedra in 1/1 and fractional as well in 2/1 AC

[c]Shell radii estimated from the respective 2/1 ACs

[d]Dominant cation sites are marked in bold

[e]Additional decorations at or near the center of the triacontahedral edges

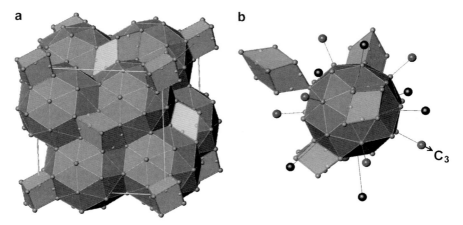

Fig. 21 a Polyhedral view of the unit cell, and **b** of the environment of a triacontahedron in the Bergman-type Mg$_{27}$Al$_{11}$Zn$_{47}$ 2/1 AC (*Pa$\bar{3}$*) in terms of triacontahedral clusters and prolate rhombohedra. Notice that both results are similar to those shown in Fig. 20 for Tsai-type 2/1 ACs. (Adapted with permission from [91]. Copyright 2006 The National Academy of Science USA)

5.3 AC Structures and QC Modeling

As discussed above, Bergman- and Tsai-type ACs appear to have the same LROs at the level of the triacontahedral clusters and beyond, with differences only in SROs within. This message suggests that Bergman- and Tsai-type i-QCs might accordingly have the same or similar structural models at the triacontahedral level and beyond. It is informative to "dig out" some hidden LRO motifs from 2/1 ACs, particularly around the pseudo fivefold axes (pseudo two- and actual threefold axes already exist). Examination of the linkages among triacontahedra and prolate rhombohedra (Figs. 20 and 22) reveals additional important clues for possible QC modeling. As shown in Fig. 22, the apex atoms on the threefold axis of the four prolate rhombohedra generate a trigonal pyramid when interconnected. Interestingly, each edge from the pyramidal apex passes nearly through the center and the two pseudo pentagonal vertices of the neighboring triacontahedra, meaning that each edge exhibits pseudo fivefold symmetry. A convenient way to measure the deviations from ideal icosahedral geometries is in terms of the angles between pairs of the three edges of the pyramid and the proper threefold axis. Table 4 lists these for some Bergman- and Tsai-type 2/1 ACs. As can be seen, they deviate only slightly from the characteristic angles of an ideal icosahedron with point symmetry $m\bar{3}5$ [99]. Such small differences indicate that the prolate rhombohedra and the trigonal pyramid so defined are coaxial with the proper threefold axis, whereas the pyramidal apex possesses pseudo icosahedral symmetry.

Further, larger prolate and oblate rhombohedra are also found to exist in 2/1 ACs, as shown in Fig. 23. The centers of the triacontahedra divide each edge by a value (1.615) close to the golden mean ($\tau = 1.618$), and thus they can be called

A Chemical Approach to the Discovery of Quasicrystals 35

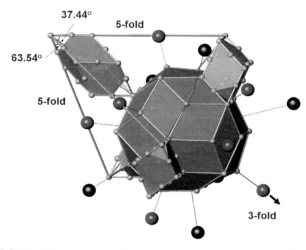

Fig. 22 The linkages between triacontahedra and prolate rhombohedra in Tsai-type 2/1 ACs (adapted from Fig. 20) afford some important clues for QC modeling. A trigonal pyramid with particular angles (as marked) and geometric features is formed when equivalent apex atoms of four prolate rhombohedra are linked. The same property exists for Bergman-type 2/1 ACs

Table 4 Angles (°) between observed three- and pseudo fivefold axes of the trigonal pyramids in 2/1 ACs

	Ideal icosahedron	Tsai-type ACs			Bergman-type ACs
		Sc–Mg–Zn	Ca–Cd	Ca–Au–In	Al–Mg–Zn
∠5–5	63.43	63.54(1)	63.53(1)	63.33(1)	63.46(2)
∠5–3	37.38	37.44(1)	37.44(1)	37.31(1)	37.39(1)
Reference	[99]	[83]	[23]	[84]	[91]

inflated prolate and oblate rhombohedra. Notice that both inflated units have the same geometric and angular relationships as listed in Table 4, and they can therefore be use as models for the decoration of QC models.

6 Remarks

The discovery of new QC/AC systems is an emerging multidisciplinary research field for chemists, and the synthetic explorations are very much open. Pseudogap tuning appears to be one of the useful routes for QC discovery, although rigid band assumptions are obviously not followed because the structures before and after tunings are naturally different. Perhaps such accomplishments really originate with similar charge distributions that are forecasted for all starting materials with a Mg_2Zn_{11}-type structures, all of which predict a pseudogap at e/a

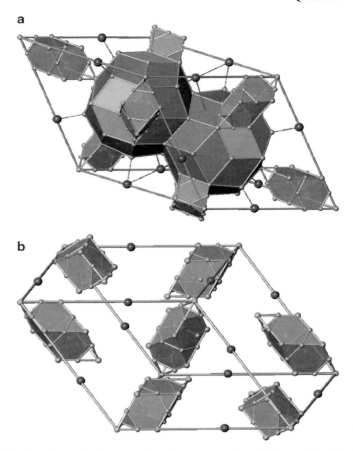

Fig. 23 Inflated prolate and oblate rhombohedra extracted from Tsai-type 2/1 ACs. The same properties exist for Bergman-type 2/1 ACs

of about 2.0. Our intuition or experiences in structure and bonding, inherited from the studies in Zintl and polar intermetallic phases, have been very informative and helpful. However, cautions remain for the pseudogap tuning concept, as it is has been confirmed mainly via synthetic explorations from Mg_2Zn_{11}-type precursors. The prediction of the QC composition in an LMTO calculation on the Ca–Au–In 1/1 AC (Fig. 16) seems noteworthy. Future endeavors to test its applicability with other structures and systems are necessary. Yet it also serves well as a *check* on the structures, symmetries, and compositions of possible precursors before pseudogap tunings. Heterometal modes of tuning via substitution of lower valent metal have also been shown to be useful in other In systems, for which e/a values are thereby lowered [100, 101]. Not every polar intermetallic with a pseudogap and empty bonding states at E_F contains what appear to be the prerequisite multiply endohedral clusters with pseudo icosahedral symmetry, and not every metal can be a component of QC/ACs. Moreover, these are just chemical guides and viewpoints, without much consideration of or insights into reaction conditions (composition, temperature, cooling rate, etc.), which are also very critical in the formation of QCs.

Nevertheless, engineering of pseudogaps has educated us and refreshed some lessons useful not only in QC searching, but also in hunting for other unknown intermetallic phases. Our results so far have also shown that we *always* encountered other novel intermetallic phases with low e/a values in the electronic tuning processes, and many more interesting and novel phases must await discovery. For example, $Mg_{35}Cu_{24}Ga_{53}$ ($e/a = 2.28$) [76], a face-centered-cubic phase featuring interpenetrating Bergman-type clusters, was obtained via pseudogap tuning of $Mg_2Cu_6Ga_5$. Similarly, $Ca_4Au_{10}In_3$ ($e/a = 1.59$) [85], a structure with novel wavy Au layers, was found in tuning reactions of $CaAu_xIn_{6-x}$. In parallel, $CaAu_3Ga$ [88] was discovered in the electronic tuning process that led to the Ca–Au–Ga i-QC. $Sc_4Mg_{0.50(2)}Cu_{14.50}Ga_{7.61(2)}$ [102], a 1/0–1/0–2/1 AC phase related to the Sc–Mg–Cu–Ga i-QC, was obtained during $Sc_{x/3}Mg_{2-x/3}Cu_6Ga_5$ searches. These discoveries have in turn helped us to improve our understanding of the structural and compositional kinships between and among polar intermetallics and QC/ACs. Polar intermetallic phases with lower e/a values have in general not been well studied, and tunings stimulated by QC searches may show us new ways to access these phases and provide us with the missing links between Hume–Rothery, polar intermetallic, and Zintl phases. The fecundity recently discovered among neighboring K–Au–(In,Sn) systems [103, 104] is an excellent hint of wonders to come in neighboring polar intermetallic systems.

Acknowledgement The research described herein was supported by the US National Science Foundation, Solid State Chemistry, via grant DMR-0444657 and earlier grants. The writing of this article was supported by Basic Energy Sciences, US Department of Energy (DOE), and all of this was carried out in the facilities of the Ames Laboratory, US-DOE.

References

1. Shechtman D, Blech I, Gratias D, Cahn JW (1984) Phys Rev Lett 53:1951
2. Janot C (ed) (1994) Quasicrystals: a primer, 2nd edn. Oxford University Press, Oxford, UK
3. Stadnik ZM (ed) (1999) Physical properties of quasicrystals. Springer, New York
4. Massalski TB, Turchi PEA eds (2005) The science of complex alloy phases. The Minerals, Metals & Materials Society, Warrendale, PA
5. Dubois JM, Janot C (eds) (2005) Useful quasicrystals. World Scientific, Singapore
6. Macia E (2000) Appl Phys Lett 77:3045
7. Kameoka S, Tanabe T, Tsai AP (2004) Catal Today 93–95:23
8. Kelton KF, Gibbons PC (1997) MRS Bull 22:71
9. Man W, Megens M, Steinhardt PJ, Chaikin PM (2005) Nature 436:993
10. Andersen BC, Bloom PD, Baikerikar KG, Sheares VV, Mallapragada SK (2002) Biomaterials 23
11. Pauling L (1985) Nature 317:512
12. Pauling L (1990) Proc Natl Acad Sci USA 87:7849
13. International Union of Crystallography (1997) Acta Crystallogr A 48:922

14. Jassen T, Chapuis G, de Boissieu M eds (2007) Aperiodic crystals from modulated phases to quasicrystals. Oxford University Press, New York
15. Goldman AI, Kelton KF (1993) Rev Mod Phys 65:213
16. Steurer W, Deloudi S (2007) Acta Crystallogr A 64:1
17. Mackay AL (1962) Acta Crystallogr 15:916
18. Bergman G, Waugh JLT, Pauling L (1957) Acta Crystallogr 10:254
19. Tsai AP, Guo JQ, Abe E, Takakura H, Sato TJ (2000) Nature 408:537
20. Sugiyama K, Kaji N, Hiraga K, Ishimasa T (1998) Z Kristallogr 213:90
21. Audier M, Pannetier J, Leblanc M, Janot C, Lang J-M, Dubost B (1988) Physica B 153:136
22. Pay Gómez C, Lidin S (2003) Phys Rev B 68:024203
23. Pay Gómez C, Lidin S (2001) Angew Chem Int Ed 40:4037
24. Lin Q, Corbett JD (2004) Inorg Chem 43:1912
25. Tsai AP (1999) In: Stadnik ZM (ed) Physical properties of quasicrystals. Springer, New York, pp 5–50
26. Steurer W, Deloudi S (2008) Acta Crystallogr A 64:1
27. Tsai AP (2003) Acc Chem Res 36:31
28. Ishimasa T (2005) In: Massalski TT, Turchi PEA (eds) The science of compex alloy phases. The Minerals, Metals & Materials Society, Warrendale, PA, p 231
29. Tsai AP (2005) In: Massalski TB, Turchi PEA (eds) The science of complex alloy phases. The Minerals, Metals & Materials Society, Warrendale, PA, p 201
30. Hume-Rothery W (1926) J Inst Metal 35:295
31. Hume-Rothery W, Raynor GV (eds) (1962) The structure of metals and alloys, 4th edn. Institute of Metals, London
32. Jones H (1937) Proc Phys Soc 49:250
33. Ishii Y, Fujiwara T (2001) Phys Rev Lett 87:206408
34. Mizutani U, Takeuchi T, Fournee V, Sato H, Banno E, Onogi T (2001) Script Metal 44:1181
35. Zeng X, Ungar G, Liu Y, Percec V, Dulcey AE, Hobbs JK (2004) Nature 428:157
36. Schaefer H, Schnering HG (1964) Angew Chem 76:833
37. Perrin C (1999) In: Braunstein P, Oro LA, Raithby PR (eds) Metal clusters in chemistry, vol. 3. Wiley-VCH, Weinheim, p 1563
38. Corbett JD (1995) J Alloys Compd 229:10
39. Svensson G, Koehler J, Simon A (1999) In: Braunstein P, Oro LA, Raithby PR (eds) Metal clusters in chemistry, vol 3. Wiley-VCH, Weinheim, p 1509
40. Corbett JD (1981) J Solid State Chem 39:56
41. Corbett JD (1981) J Solid State Chem 37:335
42. Burdett J (ed) (1995) Chemical bonding in solids. Oxford University Press, Oxford
43. Kauzlarich SM (ed) (1996) Chemistry, structure, and bonding of zintl phases and ions. VCH, New York
44. Belin C, Tillard-Charbonnel M (1993) Prog Solid State Chem 22:59
45. Miller GJ, Lee C-S, Choe W (2002) In: Meyer G, Naumann D, Wesemann L (eds) Inorganic chemistry highlights. Wiley-VCH, Weinheim, p 21
46. Corbett JD (2000) Angew Chem, Int Ed 39:670
47. Dong Z-C, Corbett JD (1996) Angew Chem Int Ed Engl 35:1006
48. Cordier G, Mueller V (1995) Z Naturforsch B 50:23
49. Tillard-Charbonnel M, Belin C (1991) J Solid State Chem 90:270
50. Shoemaker DP, Shoemaker CB (1988) In: Jari MV (ed) Introduction to quasicrystals. Academic, London, p 1
51. Corbett JD (1996) In: Kauzlarich SM (ed) Chemistry, structure, and bonding of Zintl phases and ions. VCH, New York, p 139
52. Corbett JD (1997) Struct Bonding 87:157
53. Wade K (1976) Adv Inorg Chem Radiochem 18:1
54. Dong Z-C, Corbett JD (1995) J Am Chem Soc 37:335
55. Sevov SC, Corbett JD (1991) Inorg Chem 30:4875
56. Dong Z-C, Corbett JD (1995) J Am Chem Soc 117:6447
57. Tillard-Charbonnel M, Belin C (1990) Eur J Solid State Inorg Chem 27:759

A Chemical Approach to the Discovery of Quasicrystals 39

58. Belin C, Ling RG (1982) Compt Rendus Acad Sci Ser 2 294:1083
59. Henning RW, Corbett JD (2002) J Alloy Compd 338:4
60. King RB (1996) Inorg Chim Acta 252:115
61. Villars P, Calvert LD (eds) (1991) Pearson's handbook of crystallographic data for inter-metallic phases, vol 1, 2nd edn. American Society of Metals, Materials Park, OH
62. Li B, Corbett JD (2007) Inorg Chem 45:6022
63. Palasyuk A, Dai J-C, Corbett JD (2008) Inorg Chem 47:3129
64. Cahn JW, Shechtman D, Gratias D (1986) J Mater Res 1:13
65. Beraha L, Steurer W, Perez-Mato JM (2001) Z Kristallogr 216:573
66. Andrusyak RI, Kotur BY, Zavodnik VE (1989) Sov Phys Crystallogr 34:600
67. Larson AC, Cromer DT (1971) Acta Crystallogr B 27:1875
68. Lin Q, Corbett JD (2003) Philos Mag Lett 83:755
69. Elser V (1985) Phys Rev B 32:4892
70. Markiv VY, Belyavina NN, Gavrilenko IS (1984) Izvest Akad Nauk SSSR, Met 5:227
71. Honma T, Ishimasa T (2007) Philos Mag 18–21:2721
72. Samson S (1949) Acta Chem Scand 3:809
73. Zachwieja U (1996) J Alloys Compd 235:7
74. Markiv VY, Belyavina NN (1981) Izvest Akad Nauk SSSR, Met 2:201
75. Lin Q, Corbett JD (2003) Inorg Chem 42:8762
76. Lin Q, Corbett JD (2005) Inorg Chem 44:512
77. Takeuchi T, Mizutani U (2002) J Alloy Compd 342:416
78. Ma Y, Stern EA (1998) Phys Rev B 38:3754
79. Lin Q, Corbett JD (2005) J Am Chem Soc 127:12786
80. Kaneko Y, Maezawa R, Kaneko H, Ishimasa T (2002) Philos Mag Lett 82:483
81. Samson S (1949) Acta Chem Scand 3:835
82. Lin Q, Corbett JD (2006) Philos Mag 86:607
83. Lin Q, Corbett JD (2006) J Am Chem Soc 128:13268
84. Lin Q, Corbett JD (2007) J Am Chem Soc 129:6789
85. Lin Q, Corbett JD (2007) Inorg Chem 46:8722
86. Pearson RG (1988) Inorg Chem 27:734
87. Lin Q, Corbett JD (2008) Inorg Chem 47:7651
88. Lin Q, Corbett JD (2008) Inorg Chem 47:3462
89. Henley CL, de Boissieu M, Steurer W (2006) Philos Mag 86:1131
90. Henley CL, Elser V (1986) Philos Mag B 53:L59
91. Lin Q, Corbett JD (2006) Proc Natl Acad Sci USA 103:13589
92. Takakura H, Pay Gómez C, Yamamoto Y, De Boissieu M, Tsai AP (2007) Nat Mater 6:58
93. Penrose B (1974) Bull Inst Math Appl 10:266
94. Piao S, Gomez CP, Lidin S (2006) Z Naturforsch 61B:644
95. Sun W, Lincoln FJ, Sugiyama K, Hiraga K (2000) Mater Sci Eng 294–296:327
96. Spiekermann S, Kreiner G (1998) ISIS experimental report. www.isis.rl.ac.uk/ISIS98/reports/9566.PDF. Last accessed 19 Nov 2008
97. Sugiyama K, Sun W, Hiraga K (2002) J Alloys Compd 342:139
98. Mizutani U (2005) In: Massalski TB, Turchi PEA (eds) The science of complex alloy phases. The Minerals, Metals & Materials Society, Warrendale, PA, pp 1
99. Hahn T, Klapper H (2002) In: Hahn T (ed) International tables for crystallography, vol A, 5th edn. Kluwer, Dordrecht, p 761
100. Li B, Corbett JD (2005) J Am Chem Soc 127:926
101. Li B, Corbett JD (2006) Inorg Chem 45:8958
102. Lin Q, Lidin S, Corbett JD (2008) Inorg Chem 47:1020
103. Li B, Corbett JD (2007) Inorg Chem 46:2022
104. Li B, Corbett JD (2008) Inorg Chem 47:3610

Struct Bond (2009) 133: 41–92
DOI:10.1007/430_2008_14
© Springer-Verlag Berlin Heidelberg 2009
Published online: 5 March 2009

Bonding and Electronic Structure of Phosphides, Arsenides, and Antimonides by X-Ray Photoelectron and Absorption Spectroscopies

Andrew P. Grosvenor, Ronald G. Cavell, and Arthur Mar

Abstract X-ray photoelectron spectroscopy and related techniques such as X-ray absorption spectroscopy provide useful information about the electronic structure and bonding of inorganic solids. However, interpretation of these spectra is more difficult for compounds with significant covalent bonding character. In this chapter, these spectroscopic techniques are applied to various transition-metal phosphides, arsenides, and antimonides, including a diverse class of compounds based on the MnP-type structure, as well as some binary and ternary skutterudites. Because shifts in binding and absorption energies are less pronounced, other features such as lineshape and satellite intensity become important in the determination of relative charges and valence states. These shifts can be rationalized in terms of a charge potential model and related to electronegativity differences. Valence band spectra can also be interpreted through comparison with calculated electronic structures.

Keywords: Antimonides · Arsenides · Bonding · Phosphides · X-ray absorption spectroscopy · X-ray photoelectron spectroscopy

Contents

1   Introduction 42
2   X-Ray Spectroscopies 43
    2.1   Principles of XPS 44
    2.2   XPS Instrumentation 45
    2.3   Core Level XPS Spectra 49
    2.4   XPS Valence Band Spectra 53
    2.5   Other XPS Techniques 58
    2.6   X-Ray Absorption Spectroscopy 58
3   Pnictides with the MnP-Type Structure 61
    3.1   Transition-Metal Phosphides, MP 64

A.P. Grosvenor, R.G. Cavell, and A. Mar (✉)
Department of Chemistry, University of Alberta, Edmonton, Alberta, Canada T6G 2G2
e-mail: arthur.mar@ualberta.ca

3.2	Mixed-Metal Phosphides, $M_{1-x}M'_xP$	68
3.3	Transition-Metal Arsenides, MAs, and Mixed Arsenide Phosphides, MAs$_{1-y}$P$_y$	72
4	Binary and Ternary Skutterudites	79
4.1	CoP$_3$, CoAs$_3$, and CoSb$_3$	81
4.2	REFe$_4$P$_{12}$ and REFe$_4$Sb$_{12}$	82
5	ARXPS Study of Hf(Si$_{0.5}$As$_{0.5}$)As	86
6	Conclusion	88
References		90

Abbreviations

AES	Auger electron spectroscopy
ARXPS	Angle-resolved XPS
BE	Binding energy
CHA	Concentric hemispherical analyser
cps	Counts per second
DOS	Density of states
ESCA	Electron spectroscopy for chemical analysis
EXAFS	Extended X-ray absorption fine structure
FLY	Fluorescence yield
FWHM	Full width at half maximum
IMFP	Inelastic mean free path
KE	Kinetic energy
LMTO	Linear muffin tin orbital approximation
PES	Photoemission spectroscopy
Pn	Pnicogen (e.g., P, As, Sb)
RE	Rare earth
REELS	Reflection electron energy loss spectroscopy
RSD	Residual standard deviation
SR	Synchrotron radiation
TEY	Total electron yield
UPS	Ultraviolet photoelectron spectroscopy
XAFS	X-ray absorption fine structure
XANES	X-ray absorption near-edge spectroscopy
XAS	X-ray absorption spectroscopy
XPS	X-ray photoelectron spectroscopy
XRF	X-ray fluorescence

1 Introduction

X-ray spectroscopic techniques are powerful tools for the analysis of electronic structures of materials. There are two principal techniques, X-ray photoelectron spectroscopy (XPS) and X-ray absorption spectroscopy (XAS), both of which have been used extensively for analysing many kinds of materials. In principle, both yield direct information about the electronic structure of solids, through the analysis of both core and valence states [1–3]. The two spectroscopies complement each other and offer element specificity and chemical sensitivity. As such, these experimental results provide an important confirmation of theoretical expectations derived from both simple models used in inorganic chemistry and more sophisticated band structure calculations. Despite relatively long histories, XPS and XAS have not been widely embraced by the solid-state chemistry community, in part because of the limited facilities available.

In the early development (1960–1970) of XPS, much hope was placed on the ability of the technique to probe chemical shifts that might be distinguished in the different bonding environments of atoms within solids [1, 2, 4]. Many of these studies were restricted to relatively simple solids, or those with sufficiently pronounced ionic bonding character, as found in oxides or halides, where such shifts could be more readily detected. For solids with more covalent character, the shifts were often too small to be distinguishable from the binding energies of the component elements. What has changed since that time that has compelled us to revisit the use of XPS, whose promise seemed deferred, to analyse the electronic structure and bonding of solids? Certainly there has been an improvement in instrumental capabilities of XPS spectrometers, including better resolution. While still uncommon in chemistry laboratories, they are nevertheless widely accessible in affiliated departments or support centres. Synchrotron X-ray sources have facilitated the applications of XAS and also enabled new types of XPS experiments to be conducted that have expanded the capabilities of this technique.

In this review, we present a selection of studies from our own laboratory, intended to introduce a solid-state chemist to both the practical and theoretical considerations that need to be taken into account in XPS measurements of solids with substantial covalent character. Metal phosphides, arsenides, and antimonides represent such a category of solids where the bonding retains some polarity that notions of electron counting derived from the Zintl concept still prove helpful in providing a frame of reference for comparing charge distributions. We also describe the applications of XAS to complementary studies of the electronic structure of these materials.

2 X-Ray Spectroscopies

Excellent reviews of X-ray spectroscopic techniques are available in the literature [1, 2, 5–12]. The description below summarizes the major characteristics of XPS and XAS, as applied, when relevant, to the analysis of metal phosphides, arsenides, and antimonides.

2.1 Principles of XPS

The application of XPS for analysis of solids was developed by Siegbahn and coworkers in the 1950s [4]. The technique exploits the photoelectric effect (discovered by Hertz and explained by Einstein) [2]. When an atom within a solid absorbs sufficiently energetic radiation (typically, for XPS, in the soft X-ray region, $E < 2\,\text{keV}$), an electron (called a photoelectron) is ejected [4]. In this excitation process, the energy of the absorbed X-ray is utilized to excite the electron of an inner shell orbital to the zero level of energy (the Fermi level, E_F). This excitation energy is called the *binding energy* (BE) of the electron. If the radiation energy of the incoming photon is greater than the BE, excess energy is transferred to the electron as *kinetic energy* (KE) and the electron leaves the solid to travel through vacuum with a velocity that can be measured so as to provide a relationship between the photon energy and the BE. For solids, there is also an energy required to detach the electron from the Fermi zero level into the vacuum, which is known as the *work function* (Φ_s). Conservation of energy provides the relationship given in Eq. 1, and the process [2] is diagrammed in Fig. 1:

$$h\nu = KE + BE + \Phi_s. \qquad (1)$$

To determine the BEs (Eq. 1) of different electrons in the atom by XPS, one measures the KE of the ejected electrons, knowing the excitation energy, $h\nu$, and the work function, Φ_s. Thus, the electronic structure of the solid, consisting of both localized core states (*core line spectra*) and delocalized valence states (*valence band spectra*) can be mapped. The information is element-specific, quantitative, and chemically sensitive. Core line spectra consist of discrete peaks representing orbital BE values, which depend on the chemical environment of a particular element, and whose intensity depends on the concentration of the element. Valence band spectra consist of electronic states associated with bonding interactions between the

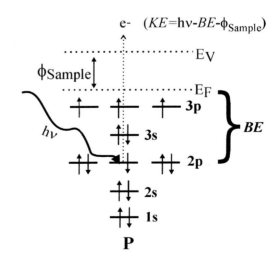

Fig. 1 Excitation of a P atom leading to ejection of a photoelectron from the 2p orbital

elements. In principle, XPS is a potentially powerful technique for detecting all elements in the periodic table (except H and He, which lack core electrons), as implied by the alternative name, *electron spectroscopy for chemical analysis* (ESCA), originally coined by its developers [4]. Because the photoelectrons that appear in the spectra emerge from only the first few layers of a solid, XPS is primarily regarded as a surface-sensitive technique [1, 5, 13]. However, when the surface is properly prepared, information about the bulk electronic structure can also be obtained.

2.2 XPS Instrumentation

Modern instrumentation has improved substantially in recent years, which has enabled the measurement of XPS spectra of superior resolution necessary to reveal the small BE shifts present in highly covalent compounds such as those studied here. In a laboratory-based photoelectron spectrometer, a radiation source generates photons that bombard the sample, ejecting photoelectrons from the surface that are transported within a vacuum chamber to a detector (Fig. 2). The vacuum chamber is required to minimize the loss of electrons by absorption in air and, if a very high quality vacuum environment is provided (as is the case with modern instruments), the surface contamination is minimized so that the properties of the bulk material are more readily determined.

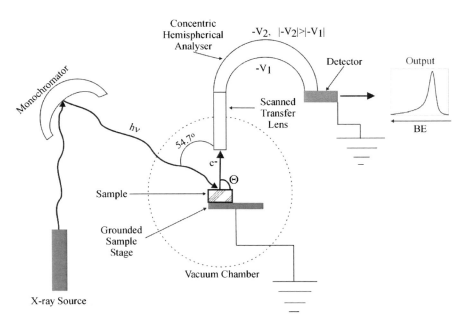

Fig. 2 Scheme of a modern XPS instrument with a monochromatic X-ray source

Common sources for laboratory XPS are Al and Mg X-rays, although He ultraviolet sources and other X-ray tubes can be used [2,6]. Alternatively, the analyser may be coupled to a synchrotron radiation source,which offers high resolution, tuneable energy for excitation [14]. The Al and Mg X-rays are typically monochromatized to remove all but $K\alpha$ radiation, improving resolution but at the expense of reduced photon flux. The detector system is ideally fixed at an angle of 54.7° to the incident photons (although this is not always the case), thereby minimizing the dependence of the excitation on orbital orientation [15, 16]. Within the vacuum chamber, the photoelectrons are collected and focused by electrostatic or magnetic fields (or both) to a concentric hemispherical analyser (CHA), where they are discriminated by energy [2]. When photoelectrons enter the aperture of the transfer lens in the CHA, their KE is reduced by a retarding electric field. The retarding energies are stepped across a range so that all photoelectrons acquire the same energy, called the *pass energy*, to enter the CHA [2]. A constant pass energy ensures constant resolution, and a low pass energy implies a high spectral resolution. Opposing electric fields on each hemisphere of the CHA (with the potential on the upper hemisphere being more negative than on the lower hemisphere) then focus the photoelectrons such that those with exactly the pass energy depart through the exit slit for detection whereas those with greater or lesser energies than the pass energy collide with the walls [2]. In the XPS instruments that we have used in this work (Kratos AXIS 165 and Kratos AXIS ULTRA), the energy-filtered photoelectrons strike a channeltron detector array, which converts the signal to an amplified pulse current and improves the signal-to-noise ratio. Scanning the retarding energy in the transfer lens results in an XPS spectrum, recorded as intensity (cps) vs. the total KE of the photoelectron (i.e., its original energy before being reduced to the pass energy) [2].

The fundamental quantity of interest, BE, is calculated from the KE (correcting for the work function Φ_s). The sample is grounded to the spectrometer to "pin" the Fermi levels to a fixed value of the spectrometer (Fig. 1) so that the applicable work function is that of the spectrometer, Φ_{sp} [2]. This instrumental parameter is a constant that can be measured. The BEs are then easily obtained from Eq. 2:

$$BE = h\nu - KE' - \Phi_{sp}, \tag{2}$$

where KE' is now the kinetic energy of photoelectrons relative to the spectrometer [1].

Proper sample preparation is essential for obtaining meaningful information about electronic structure from an XPS spectrum. Any solid exposed to the atmosphere will acquire a few layers of surface oxides or contaminants that must be removed. Fresh surfaces of single-crystal or polycrystalline samples can be obtained by cleaving in vacuum, provided that the specimens are larger than the spot size of the source ($\sim 400 \times 700\,\mu m^2$ in Kratos spectrometers). Small crystals can also be ground under an inert atmosphere in a glove-box directly attached to the vacuum chamber. A widely used technique is to sputter the surface with a beam of energetic Ar^+ ions in vacuum to remove contaminant or oxide layers (Fig. 3) [2]. However, deleterious side effects may occur that modify the composition of the surface, such as the preferential removal of lighter elements, introduction of defects,

Fig. 3 Ar⁺ ion sputtering of a surface, which removes contaminant or oxide layers, exposing a clean surface for analysis

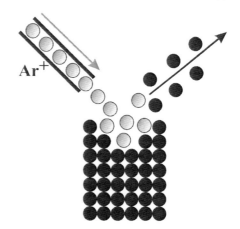

or reduction of atoms (commonly observed when ionic oxides such as CeO_2 are sputtered) [2, 17, 18]. Annealing the specimen at high temperature after sputtering helps heal the surface, but this is not always appropriate, depending on the sample. It is important to collect spectra of the sample before and after sputtering to ascertain if the surface has been adversely affected.

As the sample actively loses electrons during the XPS measurement, it must have proper electrical ground contact with the spectrometer and sample holder stage to prevent charge build-up on the specimen. For well-conducting samples (metals and semiconductors), charging presents no serious problems as current readily passes from the spectrometer to the sample, neutralizing the charge. For poorly conducting samples (insulators and large band gap semiconductors), charging cannot be completely eliminated by grounding and the effect is to raise the apparent BE of all photoelectron peaks [1, 2, 5]. In these cases, the BE values can be corrected by readjusting the measured value to a reference value [1, 2, 5]. The 1s signal from adventitious carbon (desorbed from the walls of the vacuum chamber) found on all surfaces can be used; readjusting all BE values to the established C1s standard value of 284.8 ± 0.2 eV provides true BE values [1,2,5]. For insulating samples, charging is often so severe that the resulting spectrum becomes very broad and nearly impossible to interpret. Charge neutralization may be achieved in such situations by bathing the surface with low-energy (low voltage) electrons emitted from a tungsten filament, compensating for the charging and allowing well-resolved spectra to be obtained [1, 2, 5].

Although most of the results described in this review are aimed towards an understanding of the electronic structure of the bulk solid, the most widespread applications of XPS exploit its surface sensitivity, as a result of the numerous interactions that an ejected photoelectron undergoes as it travels through the surface layers [2]. The photoelectron collides with other electrons, either elastically so that its trajectory changes, or inelastically so that its KE decreases (by 10–40 eV per collision) [2, 19, 20]. The distance travelled between inelastic collisions is called the *inelastic mean free path* (IMFP, λ), and depends on the sample and the KE of the exiting photoelectron [21]. Photoelectrons originating from deep within the solid undergo more scattering events, losing enough KE that they are not likely to escape to the surface and be detected [19]. The sampling depth is maximized when the KE of the photoelectron is highest and when the *take-off angle* Θ (between the surface and detector) is $90°$ [2]. In particular, an XPS peak contains information from a depth $d = 3\lambda \sin \Theta$, meaning that a photoelectron can travel a maximum distance of 3λ (when Θ is $90°$) before it becomes $>95\%$ probable that it will undergo an inelastic collision and lose energy [1, 2]. With typical λ values of $10 - 30\,\text{Å}$ [21], a maximum depth of 3–9 nm can be sampled. Those photoelectrons that have lost energy but manage to breach the surface make up the intense stepped background seen in all XPS spectra [1, 2]. Reducing the take-off angle increases the amount of material through which photoelectrons must travel before escaping, enhancing the probability that they will undergo inelastic scattering. This dependence on take-off angle forms the basis for *angle-resolved XPS* (ARXPS) [1, 13], which reveals useful information about surface structure, as we demonstrate with an example on the oxidation behaviour of an arsenide in Sect. 5.

Since their inception, X-ray photoelectron spectrometers have undergone numerous improvements in their source and detector systems. Early on, the advantages of monochromatic X-ray sources were already well recognized, providing reduced spectral lineshapes and thereby higher spectral resolution [1]. To increase the flux of photoelectrons entering the CHA, the lens system has been updated to include not only electrostatic but also magnetic focus [1]. Such a combined lens system allows more intense spectra to be collected, increasing the signal-to-noise ratio and improving the detection limit. Detector technology has also advanced rapidly. The spectra presented herein were collected with a channeltron detector system, which contains eight detection channels. New XPS systems have recently been developed that contain a multi-channel plate stack, with the number of detection channels increased to 128 [22]. Not only do such spectrometers show much improved signal-to-noise ratios, but they also allow high-resolution spectra to be collected very quickly [22], which would be advantageous in dynamic studies (e.g., when surface degradation occurs over time upon exposure to X-ray beams). A new development is to use XPS instruments to perform 2D imaging experiments, by adding a third position-sensitive hemisphere to the CHA and moving the sample stage in small precise steps with calibrated motors [1,5]. Exploiting improved energy resolution and a reduction of beam size (from thousands or hundreds to tens of micrometres in diameter) allows surfaces to be mapped elementally or chemically (i.e., of the same element but in different oxidation states) [1].

Compared to metal anode X-ray sources, synchrotron radiation (SR) provides unmatched intensities. In addition, high resolution photon energies, which can be provided by monochromatized synchrotron radiation, allow the examination of deeper core levels of, for example, transition metals (with 1s energies $>5\,\text{keV}$), thereby extending the scope of XPS [14]. Although some high energy XPS work has been done with high energy metal anode X-ray sources (e.g., Cu X-rays) [23], the much broader linewidths of such sealed tube X-ray sources hamper the chemical shift analysis – a deficiency that is overcome with SR. Moreover, the tunability of the photon energies accessible with synchrotron radiation allows new types of analyses, such as variable energy photoemission spectroscopy (PES), to be performed (Sect. 2.5).

2.3 Core Level XPS Spectra

2.3.1 Survey vs. High-Resolution Spectra

A typical XPS experiment begins with a *wide-scan* or *survey* spectrum, as shown for Hf metal in Fig. 4. Such spectra are collected with a high pass energy (reduced resolution), a large energy step (0.7 eV), and a wide energy envelope (>1000 eV) to reveal all peaks that can be excited with the X-ray source at once, superimposed on the stepped background arising from inelastic scattering. Because different elements possess characteristic signatures for their binding energies, survey spectra are useful for qualitative chemical analysis of a surface. The concentration of different elements can then be obtained from the integrated peak intensities I (after the energy loss background is subtracted), modified by several correction factors:

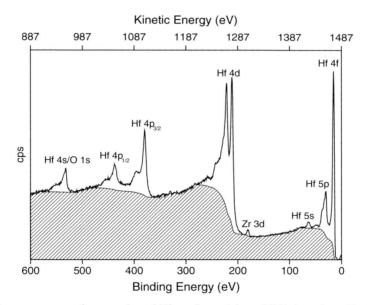

Fig. 4 Survey spectrum of sputter-cleaned Hf metal containing ∼3% Zr impurity, with all visible core-line peaks labelled, collected using a monochromatic Al Kα X-ray source (1486.7 eV). The stepped background (*hatched*), found in all XPS spectra, arises from photoelectrons that lose KE by inelastic electron scattering as they travel through the surface and into vacuum

cross-section (σ), spectrometer factors (K), and IMFP (λ) [10]. The cross-section, σ, is the probability that a photoelectron will be excited from a particular orbital. Photoelectron cross-sections have been calculated for most elements in the periodic table for a range of excitation energies. Spectrometer factors include the efficiency of the detector, the presence of stray magnetic fields (which affect the transmission of low-energy electrons), and the transmission function of the instrument [10]. Corrections for the IMFP, defined above, take into account that many photoelectrons lose kinetic energy because of inelastic collisions so that the intensity of the peak, having a defined BE width, is lower than expected [19]. The percent concentration of a given element i is then obtained from:

$$C_i = \frac{I_i/(\sigma_i K_i \lambda_i)}{\sum_{j=1}^{n} I_j/(\sigma_j K_j \lambda_j)} \tag{3}$$

where the numerator is the corrected intensity of the species under consideration and the denominator is the sum of the corrected intensities for all species [10]. The detection limit for XPS analysis is typically on the order of 0.1–1.0 atomic % [1].

The most common application of XPS, however, takes advantage of its sensitivity to different chemical environments around a given atom. Subtle changes in the photoelectron peaks can be revealed in *high-resolution* spectra, which are collected with a low pass energy, a small energy step ($\sim$0.1 eV), and a narrow energy envelope (<100 eV). The phosphorus 2p spectrum in FeP illustrates some of the features that can be interpreted in a high-resolution spectrum (Fig. 5), including its splitting into a doublet, the BE for these peaks, their lineshapes, and their intensities.

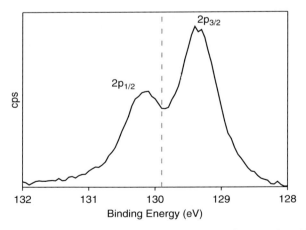

Fig. 5 High-resolution P 2p XPS spectrum for FeP, with its two distinct spin–orbit coupled final states. (The *dashed vertical line* indicates the $2p_{3/2}$ BE for elemental P)

2.3.2 Initial and Final State Energies

Previously, the BE was defined as the energy required to ionize an electron from its ground state orbital to the Fermi level:

$$M \xrightarrow{h\nu} M^+ + e^- \qquad (4)$$

Because the number of electrons (n) in the atom in its initial ground state (M) is different from the number of electrons ($n-1$) in its final excited state (M^+), the BE cannot be directly equated with the orbital energy. Properly defined, the BE is actually the energy difference between the initial and final states of the atom [2, 24]. Early in the history of XPS, it was assumed that the final state could be represented by a single wavefunction (with a single energy) and that no relaxation occurs (Koopmans' theorem) [25]. This approximation that the BE equals the orbital energy from which the electron originated provided a satisfactory interpretation of early spectra whose resolution was too low to reveal fine structure in the photoelectron peaks. Because the photoionization event leaves a positively charged hole in the core of the atom, the outer electrons are effectively less well-screened from the nuclear charge and experience a greater attraction to the nucleus [1, 5]. Various relaxation mechanisms can take place such that several final states may be possible, leading to the observation of several photoelectron peaks instead of just one. As well, the final excited state can decay through an interactive process called *Auger emission* wherein an electron in an outer orbital drops down to fill the core hole and gives the released energy to another electron, which departs from the atom [2]. Although Auger processes can be observed in XPS spectra, they are examined by another technique, Auger electron spectroscopy (AES), which utilizes a source of electrons instead of X-rays [1, 2]. The excited core hole of the atom can also be filled by an outer orbital electron and the energy released as an X-ray photon. This process is X-ray fluorescence (XRF) and is used (see below) as a detection tool.

2.3.3 XPS Peak Splittings

Excitation from an orbital with non-zero orbital angular momentum (i.e., p, d, or f orbital) results in two peaks that originate from *spin–orbit coupling* [2]. After removal of the photoelectron in a core orbital, the remaining electron can be spin-up or spin-down. The vector coupling of the spin angular momentum (s) of this electron with the orbital angular momentum (l) then leads to two possible final states with total angular momentum $j_1 = l + s$ or $j_2 = l - s$ [2]. The peak arising from the spin-up case is always lower in BE than from the spin-down case, and their intensities are in a ratio proportional to the degeneracies of these states, $(2j_1 + 1)/(2j_2 + 1)$ [2, 26]. For example, the phosphorus 2p spectrum of FeP (Fig. 5) reveals a doublet consisting of a $2p_{3/2}$ peak at lower BE and a $2p_{1/2}$ peak at higher BE, with intensity ratio 2:1.

Transition-metal and rare-earth atoms that contain partially occupied d or f valence subshells also give rise to spectral fine structure, often with very complicated *multiplet splitting* [2, 27, 28]. The spin-unpaired valence d or f electrons can undergo spin–orbit coupling with the unpaired core electron (remaining in the orbital from which the photoelectron was removed), producing multiple non-degenerate final states manifested by broad photoelectron peaks [2, 27].

2.3.4 XPS BE Shifts

With the caveat that the BE is affected by many factors besides the orbital energy, BE shifts provide useful information about the chemical environment around an atom within a solid. In general, BE shifts provide a measure of how well an electron is screened (by other electrons) from the nuclear charge on an atom, from which conclusions about bonding character (degree of electron transfer to or from surrounding atoms) and oxidation state can be inferred [2, 4]. The greater the screening, the easier it is to remove a photoelectron and the lower the BE. Thus, relative to the neutral element, photoelectron peaks have higher BE in cations and lower BE in anions. For solid-state chemists accustomed to dealing with band structures where orbital energies are usually reported as negative values, it may be helpful to keep in mind that higher BE corresponds to lower energy levels (tightly bound electrons; i.e., cationic state) and lower BE corresponds to higher energy levels (loosely bound electrons; i.e., anionic state). Figure 6 shows that Fe $2p_{3/2}$ BE values increase, as expected, with higher oxidation state in various oxides and halides

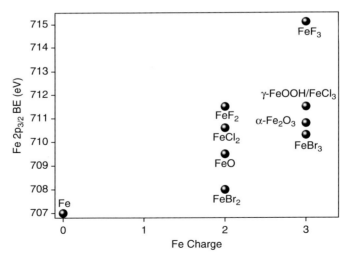

Fig. 6 Plot of Fe $2p_{3/2}$ BE vs. oxidation state for iron-containing compounds. The BE also varies between compounds having the same Fe oxidation state because of changes in the environment and screening of the nuclear charge provided by the ligands

Bonding and Electronic Structure of Phosphides

compared to elemental iron [29]. Although *intraatomic* effects provided by the change in charge (Fe^0, Fe^{2+}, Fe^{3+}) are generally the most important, *interatomic* effects provided by the neighbouring atoms can also be significant and are responsible for additional BE shifts when the ligand is changed around the iron centre [29]. While the explanation of some of these BE shifts can be framed in terms of an electronegativity difference between metal and ligand, leading to a polarization of electron density within bonds, a proper analysis of these interatomic effects includes the effect of all the surrounding atoms (which generates an opposing *Madelung potential*) [1, 5].

2.3.5 XPS Satellites

Other more complicated types of relaxation mechanisms can occur to stabilize the final state after photoionization, generating additional satellite peaks. Because the final state is always lower in energy than that of the idealized "frozen-core" ion, these satellite peaks correspond to electrons that escape with lower KE and are thus located at apparently higher BE than the main peak. *Plasmon loss* peaks arise when photoelectrons interact with, and lose some of their energy to, delocalized valence band electrons, which undergo collective excitations called plasmons [30]. These satellite peaks appear as small oscillations, often extending to several electron volts to higher BE beyond the main peak. *Shake-up* peaks occur when the final state relaxes by excitation of a valence electron to an empty orbital higher in energy [1, 31–33]. While such peaks may seem to be a nuisance in the interpretation of XPS spectra, they can provide additional information about bonding. In particular, their intensities are influenced by the degree of charge transfer to or from surrounding atoms and also by the type of orbitals that are involved [31–33]. When shake-up processes take place that promote electrons into an empty band where there is now a continuum of states, as is present in a metallic solid, the effect is to produce an asymmetric lineshape broadened at higher BE to the main peak [1, 34].

2.4 XPS Valence Band Spectra

The low BE region of XPS spectra ($<20 - 30\,eV$) represents delocalized electronic states involved in bonding interactions [7]. Although UV radiation interacts more strongly (greater cross-section because of the similarity of its energy with the ionization threshold) with these states to produce photoelectrons, the valence band spectra measured by ultraviolet photoelectron spectroscopy (UPS) can be complicated to interpret [1]. Moreover, there has always been the concern that valence band spectra obtained from UPS are not representative of the bulk solid because it is believed that low KE photoelectrons have a short IMFP compared to high KE photoelectrons and are therefore more surface-sensitive [1]. Despite their weaker intensities, valence band spectra are often obtained by XPS instead of UPS because they provide valuable information about electronic structure that can be more easily compared with the density of states obtained by band structure calculations [1].

Surprisingly little has been done to take advantage of these valence band spectra, perhaps because of some of the challenges in interpretation. In principle, it should be possible to fit these spectra with component peaks that correspond to contributions from individual atomic valence orbitals. However, a proper comparison of the experimental and calculated band structures must take into account various correction factors, the most important being the different photoelectron cross-sections for the orbital components.

As an illustration of this fitting procedure, a moderately challenging example is treated here, showing how knowledge of the crystal structure, calculation of the band structure, and judicious choice of XPS fitting parameters can allow useful information about bonding to be extracted from a valence band spectrum. $Hf(Si_{0.5}As_{0.5})$As is a new ternary arsenide adopting the well-known ZrSiS-type (or PbFCl-type) structure [35]. The structure itself, common to many other compounds of the formulation MAB, is relatively simple: it consists of a stacking of square nets, with the nets containing the A atoms (disordered Si/As1) being twice as dense as those containing the M (Hf) or B (As2) atoms (Fig. 7) [36, 37]. What is difficult is how to make sense of the bonding. Interestingly, the extreme ionic viewpoint implicit in the charge-balanced Zintl formulation, $Hf^{4+}(Si_{0.5}As_{0.5})^{1-}As^{3-}$ accounts handily for (i) the intermediate A–A separations

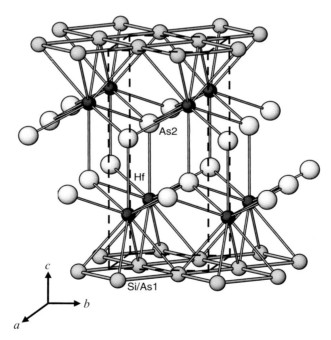

Fig. 7 Structure of $Hf(Si_{0.5}As_{0.5})$As. Reprinted with permission from [35]. Copyright Elsevier

Bonding and Electronic Structure of Phosphides

(2.5746(4) Å) within the A net, implying the presence of weak anion–anion bonding associated with not fully reduced Si or As atoms, and (ii) the long B–B separations (3.6410(5) Å) within the B net, implying the presence of "isolated" (fully reduced) As^{3-} anions [35]. However, such formulations clearly overemphasize the degree of charge transfer that takes place, especially given the small electronegativity difference between Hf and Si or As ($\Delta\chi_{Si-Hf} = 0.5$; $\Delta\chi_{As-Hf} = 0.7$) [38]. If Pauling's correlation of electronegativity difference with bonding character is assumed, the Hf–Si, Hf–As, and Si–As bonds are expected to contain less than 20% ionic character [38]. Experimental high-resolution core-level spectra reveal that the BEs of the Si $2p_{3/2}$ (98.9 eV) and As $3d_{5/2}$ (40.8 eV) peaks in Hf($Si_{0.5}As_{0.5}$)As are lower than in elemental Si (99.5(3) eV) and As (41.7(2) eV), respectively, indicative of anionic Si and As species [35, 39]. Moreover, there is only one set of As 3d peaks, so that the supposed distinction between As atoms in the A vs. B nets is not seen [35]. If a linear correlation between BE and nominal oxidation states is assumed for other Si- and As-containing compounds whose XPS spectra have been measured, these BEs suggest charges of Si^{1-} and As^{1-} in Hf($Si_{0.5}As_{0.5}$)As [35]. In contrast, the Hf $4f_{7/2}$ BE is higher in Hf($Si_{0.5}As_{0.5}$)As (14.8 eV) than in elemental Hf (14.2 eV), consistent with cationic Hf species, but nowhere near as oxidized as expected [35, 39]. For comparison, the Hf $4f_{7/2}$ BE is 16.7 eV in HfO_2 [35, 39].

With the core-line BE shift analysis as a start, the valence band spectrum can be examined to yield more insight (Fig. 8a). When valence band spectra are well-resolved, as is the case here, it becomes feasible to attempt a deconvolution into components to separate individual valence states. The first task is to compare qualitatively this spectrum with results from band structure calculations. Three models were considered, differing in the occupancy of the disordered A site and the type of bonding interactions present in the denser square net: (i) $HfAs_2$ (As–As bonding only), (ii) HfSiAs (Si–Si bonding only), and (iii) ordered Hf($Si_{0.5}As_{0.5}$)As (Si–As bonding only) (Fig. 8b). The intensities of the peaks in the valence band spectrum are not directly comparable to the DOS curves because of variations in the photoionization cross-sections of different elements [1]. Incidentally, the sharp cutoff at 0 eV in the valence band spectrum is characteristic of metallic behaviour, confirmed by partly filled bands being crossed at the Fermi level in the DOS curves and by temperature-dependent resistivity measurements [35]. The shoulder at the Fermi edge corresponds to filled Hf 5d states; their occupation suggests that Hf is indeed not fully oxidized, arguing against an unrealistic Hf^{4+} species. The intense broad peak centred around 3 eV represents As 4p and Si 3p states. Examination of the DOS curves shows that the As 4p states should be at slightly lower BE for the As2 atoms than for the As1 atoms. If only Si–As bonding is active in the denser square net, the Si 3s states lie higher in energy (lower BE) than the As2 4s states, but introduction of Si–Si bonding shifts the Si 3s states to lower energy (higher BE) so that they overlap the As2 4s states [35]. Thus, the peak centred at 9 eV corresponds to both Si 3s and As2 4s states. Finally, the peak at 12 eV is unambiguously assigned as As1 4s states.

With these assignments, it is possible to attempt a quantitative fitting of the valence band spectrum. The parameters involved include lineshape, BE, FWHM, and

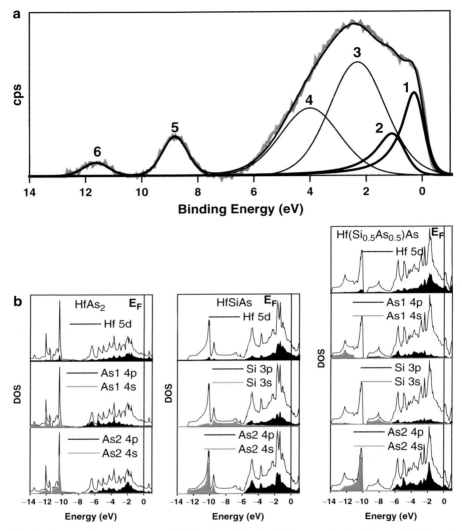

Fig. 8 a Valence band spectrum of Hf(Si$_{0.5}$As$_{0.5}$)As. **b** DOS curves obtained from LMTO band structure calculations on different ordered models for Hf(Si$_{0.5}$As$_{0.5}$)As [119]. Reprinted with permission from [35]. Copyright Elsevier

peak area of each component peak, which can be adjusted in the CasaXPS software package to optimize the fit, subject to user-defined constraints [40]. The goodness of fit is gauged by the residual standard deviation (RSD) for the difference between the simulated and experimental spectra, with an RSD value below 2.0 judged to be acceptable. A preliminary fitting is first performed by inspection, and then the component peak widths and energies are constrained such that they can only shift from

Bonding and Electronic Structure of Phosphides

their initial values by $2\,\mathrm{eV}$ or less. Convergence is reached when the RSD shifts are less than ± 0.1 in successive cycles of refinement. In Fig. 8a, the lineshapes of component peaks were fit with a combined Gaussian (70%) and Lorentzian (30%) profile (representing spectrometer and sample effects, respectively), to match those of peaks seen in the core-line spectra. Peaks 1 and 2 represent the Hf $5d_{5/2}$ and $5d_{3/2}$ spin–orbit doublet, fitted with asymmetric lineshapes similar to the profile found in the Hf 4f core-line spectrum. The peak splitting was fixed at $\sim 1\,\mathrm{eV}$, on the basis of additional information from photoemission spectra [35]. Peak 3 is the As2 4p component, and peak 4 combines the Si 3p and As1 4p components in a 1:1 ratio. In principle, these p components should be split into spin–orbit doublets ($np_{3/2}$ and $np_{1/2}$), but the separation is small and the fitting is simplified by using single peaks. Peak 3 is more intense than peak 4 because the cross-section for As 4p is larger than for Si 3p under the excitation energy (Al Kα X-rays) used [41,42]. The ratio of As2 to As1 atoms (2:1) is also taken into account. Peak 5 represents the As2 4s and Si 3s states (in a 2:1 ratio), and peak 6 represents the As1 4s state.

The integrated intensities of the fitted component peaks should then be related to the electron population of different valence states, subject to correction factors, according to the same equation used earlier for quantitative analysis of survey XPS spectra (Eq. 3) [10]. Because photoelectron KEs are similar throughout the valence band region, spectrometer-dependent factors and IMFP values can be assumed to be the same for all states, so that the equation simplifies to:

$$C_i = \frac{I_i/(\sigma_i)}{\sum\limits_{j=1}^{n} I_j/(\sigma_j)} \tag{5}$$

When corrections are applied using two different sets of photoionization cross-sections available in the literature on two separate samples of Hf(Si$_{0.5}$As$_{0.5}$)As, electron populations can be extracted, as listed in Table 1. The results suggest a

Table 1 Electron population of valence states in Hf(Si$_{0.5}$As$_{0.5}$)As[a]

State	Sample 1		Sample 2	
	[42][b]	[41][b]	[42][b]	[41][b]
Hf 5d (peaks 1, 2)	1.5 [18.2]	1.8	1.9 [23.1]	2.3
As2 4p (peak 3)	5.1 [44.4]	5.2	4.8 [41.6]	4.9
As1 4p/Si 3p (peak 4)	5.8 [28.2]	6.0	5.8 [27.6]	5.8
As2 4s/Si 3s (peak 5)	0.9 [6.9]	0.4	0.8 [6.1]	0.4
As1 4s (peak 6)	0.2 [2.3]	0.1	0.2 [1.5]	0.1
Total Hf charge	2.5+	2.2+	2.1+	1.7+
Total anion charge [(Si$_{0.5}$As$_{0.5}$)As]	2.5−	2.2−	2.1−	1.7−
Average charges	2.1+ [Hf], 2.1− [(Si$_{0.5}$As$_{0.5}$)As]			

[a]Calculated using Eq. 5 and a total valence electron population of $13.5\,\mathrm{e}^-$. The total peak areas of the peaks used to fit different states in the spectra are listed in square brackets
[b]Reference for cross-section (σ)

charge formulation Hf^{2+} $[(Si_{0.5}As_{0.5})As]^{2-}$, the reduced magnitude of the charges (relative to an ionic picture) reflecting the substantial covalent bonding character and the strong hybridization of valence states from different elements [35]. Moreover, the chemically inequivalent As1 and As2 atoms are indistinguishable from their charges, both being 1– [35].

2.5 Other XPS Techniques

One of the most exciting developments in modern X-ray spectroscopy is the now widespread availability of synchrotron radiation sources. By virtue of its much higher intensity and the tunability of its wavelength over a broad range, synchrotron radiation permits more sophisticated experiments to be performed [43].

2.5.1 Photoemission Spectroscopy

Instead of a fixed-energy X-ray source, as used in conventional laboratory-based XPS, photoemission spectroscopy (PES) takes advantage of the ability to select a range of excitation energies with synchrotron radiation. With the BE of a photoelectron being fixed, varying the excitation energy modifies the KE, IMFP, and cross-section for a photoelectron emerging from a given orbital [13,42,44]. The intensities of photoelectron peaks will thus depend on the excitation energy. Monitoring how the intensity of a valence band spectrum changes with excitation energy provides information about the atomic orbital contributions. A good example of this type of experiment is seen again with $Hf(Si_{0.5}As_{0.5})As$, where increasing the excitation energy from 250 to 500 eV enhances the photoionization cross-section for Hf 5d states (relative to the As 4p and Si 3p states) (Fig. 9a) [42]. The valence band spectra then reveal that the Hf 5d states contribute to the region between 4 and 0 eV, where an increase in intensity is observed at higher excitation energies (Fig. 9b) [35]. The feature nearest the Fermi edge is assigned as the Hf $5d_{5/2}$ peak, and that at higher BE to be the $5d_{3/2}$ peak [35]. This assignment can be confirmed by a related type of experiment called resonant photoemission spectroscopy, in which specific valence states can be directly probed by inducing transitions from a core level to a partially occupied band [35].

2.6 X-Ray Absorption Spectroscopy

The most prevalent technique exploiting synchrotron radiation is X-ray absorption spectroscopy (XAS, also called X-ray absorption fine structure, XAFS). Two related types of experiments are conducted: X-ray absorption near-edge spectroscopy (XANES), which probes the initial absorption edge and related nearby structure, and

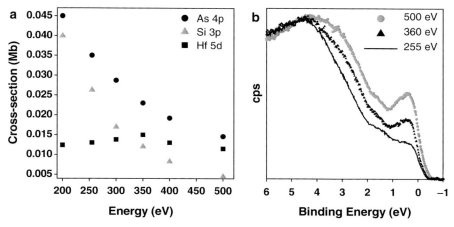

Fig. 9 a Variation in photoionization cross-sections [42]. **b** PES valence band spectra for Hf(Si$_{0.5}$As$_{0.5}$)As at three different excitation energies, normalized to the As2 4p peak. Reprinted with permission from [35]. Copyright Elsevier

extended X-ray absorption fine structure (EXAFS) [45]. XANES provides chemical sensitivity and element specificity. EXAFS, wherein the long absorption profile following the edge is measured, yields a profile of interaction of the ejected electrons with the surrounding atoms, which can be processed to reveal the local structure (number and distances of neighbouring atoms) surrounding the excited atom.

As the energy of the impinging X-rays increases, the sample will absorb radiation (according to $I = I_o e^{-\mu t}$, where μ is the absorption coefficient and t is the sample thickness) when it exceeds the threshold energy required to excite a core electron into unoccupied bound (bonding) or continuum states [46–48]. This sharp increase in absorption probability appears as an "edge" in a XANES spectrum. While the XANES spectrum itself can be considered to be a map of the empty electronic states, the edge corresponds to the energy of a filled state. Like XPS peaks in core-line spectra, XANES absorption edges exhibit energy shifts and lineshapes that are characteristic of the chemical environment around a given atom [3]. An example of the Co K-edge spectrum in CoP is shown in Fig. 10. The absorption process normally follows dipole selection rules in which $\Delta l = \pm 1$ (e.g., Co 1s → Co 4p or P 3p) [3, 45]. Below the absorption threshold, pre-edge features are also often observed, such as those resulting from weaker quadrupolar excitations (e.g., Co 1s → Co 3d), and can provide information about coordination environment and charge [3]. In contrast to XPS, which probes all of the energetically accessible orbital electrons in an atom, XANES, which is the process of exciting deeper orbital electrons into unoccupied (but still potentially binding) levels, provides an excitation spectrum that also probes these unoccupied orbital states, which can be accessed by the dipole excitation process. The energy selectivity of the synchrotron radiation allows orbitals of different angular momentum character (e.g., s, p, d) to be excited, thus allowing the electronic structure of the system to be mapped.

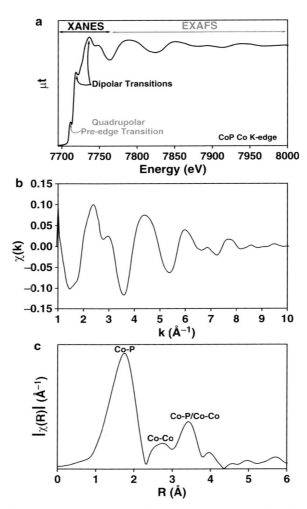

Fig. 10 **a** Co K-edge XAS spectrum for CoP collected in transmission mode, showing the approximate regions where XANES and EXAFS features are observed and the assignment of dipolar and quadrupolar transitions. **b** EXAFS (χ) vs. k curve. **c** Fourier transform of EXAFS

The shifts in the precise value of the edge reflect the chemical valence of the element in a fashion similar to the BE shifts observed in XPS in that they arise from differences in the screening of the probed electron by other electrons, and similar charge and potential models can be applied. In general, anionic states have lower absorption edge values and cationic states, higher edge energies. Because dipole selection rules operate, certain geometries create characteristic features. For example, tetrahedral environments have allowed s-to-d transitions, which are "forbidden" (and therefore of lower intensity) in regular octahedral environments [3]. We will show below that XANES and XPS yield highly complementary information.

Bonding and Electronic Structure of Phosphides

The EXAFS part of the spectrum arises beyond the absorption edge. The exact boundary between the two sections of the absorption spectrum is indeterminate and depends on the system. The EXAFS pattern is created when the photoionized core electrons (as with the XPS experiment) travel through and out of the surface of the solid. Some of them interact with neighbouring atoms and become backscattered towards the originating atom [47]. Treated as waves ($\lambda = 2\pi/k$), the backscattered electrons interfere constructively or destructively with outgoing photoelectrons to produce an interference pattern manifested as oscillations, called EXAFS (extended X-ray absorption fine structure), above the edge in the absorption spectrum [47]. Analysis of these EXAFS features (not discussed here) provides information about interatomic distances and coordination environments surrounding the excited atom, and is especially valuable when normal crystallographic methods are not applicable [45, 47].

An XAS experiment is performed by focusing X-rays with different energies selected by a monochromator onto a sample, and comparing the incident (I_o) and transmitted intensities (I) measured with inert-gas-filled (N_2 or He) ionization current detectors [46, 49]. A spectrum of absorption intensity $\mu t = -\ln(I/I_o)$ vs. excitation energy is then obtained. Alternatively, the absorption spectrum can be obtained indirectly by measuring the X-ray fluorescence (XRF) intensity (I_f) that is produced when inner shell orbital vacancies decay to a more stable state [46]. In this fluorescence experiment, the absorption coefficient μ is replaced by the ratio I_f/I_o [46, 49]. For hard X-rays, detection of the normal absorption generally suffices (but both absorption and fluorescence data can be collected simultaneously), whereas for soft X-rays (e.g., $<5\,\mathrm{keV}$), detection of the fluorescence yield (FLY) is often preferred because the emerging radiation is not well transmitted except through the thinnest samples. Absorption-type spectra can also be collected by detecting the emitted Auger and secondary electrons, as well as photoelectrons, produced by the excitation [50]. This type of detection is performed by electrically connecting the sample to an insulated holder and measuring the current flowing from an external ground source through an ammeter. This is referred to as total electron yield (TEY) and is considered to be surface-sensitive compared to FLY detection because the escaping (low energy) electrons are primarily those which are in the vicinity of the surface.

3 Pnictides with the MnP-Type Structure

Among binary transition-metal pnictides, only the first-row transition-metal phosphides have been analysed by XPS extensively, whereas arsenides and antimonides have been barely studied [51–61]. Table 2 reveals some general trends in the P $2p_{3/2}$ BEs for various first-row transition-metal monophosphides, as well as some metal- and phosphorus-rich members forming for a given transition metal. Deviations of as much as a few tenths of an electron volt are seen in the BEs for some compounds measured multiple times by different investigators (e.g., MnP), but these

Table 2 P $2p_{3/2}$ BEs for various binary first-row transition-metal phosphides

Compound[a]	P $2p_{3/2}$ BE (eV)	Reference
P (0)	130.0	[39]
	130.2	[39]
	129.9	[52]
CuP_2 (0.31)	129.7	[39]
	129.6	[39]
Ni_5P_4 (0.31)	129.5	[39]
Ni_2P (0.31)	129.5	[39]
CoP (0.36)	129.5	[39]
	129.5	[58]
Co_2P (0.36)	129.4	[39]
FeP_3 (0.42)	129.7	[39]
	129.6	[39]
FeP_2 (0.42)	129.8	[39]
FeP (0.42)	129.5	[39]
	129.5	[52]
	129.3	[58]
Fe_2P (0.42)	129.5	[39]
Fe_3P (0.42)	129.4	[39]
MnP (0.46)	129.3	[39]
	129.4	[52]
	129.2	[58]
CrP_3 (0.50)	129.1	[39]
CrP (0.50)	129.6	[39]
	129.1	[58]
VP (0.61)	129.1	[52]
	129.1	[58]
TiP (0.74)	128.4	[39]
	128.4	[52]
	128.5	[58]
ScP (0.86)	127.7	[52]

[a]Values for $\Delta\chi_{P-M}$ are shown in parentheses [62]

are probably related to differences in calibration procedures and instrument resolution. If a more typical BE precision of ± 0.1 eV is accepted, it can be seen that the P $2p_{3/2}$ BE in the monophosphides decreases roughly linearly as the difference in electronegativity $\Delta\chi_{P-M}$ increases, with the early transition-metal phosphides experiencing the most prominent BE shifts (relative to elemental P). For the other compounds, BE shifts tend to be greater with a change in the identity of M than in P concentration.

Our focus is on the most comprehensively studied series, the monophosphides of the first-row transition metals, whose structures successively distort from NaCl-type (ScP) to TiAs-type (TiP), NiAs-type (VP), MnP-type (CrP, MnP, FeP, CoP), and NiP-type, forming stronger metal–metal and phosphorus–phosphorus bonding with greater electron count (Fig. 11) [63–65]. The P atoms are six-coordinate, but

Bonding and Electronic Structure of Phosphides

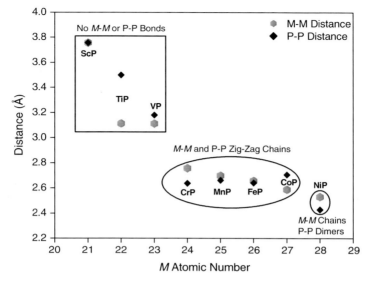

Fig. 11 Plot of P–P and *M*–*M* distances for first-row transition-metal monophosphides

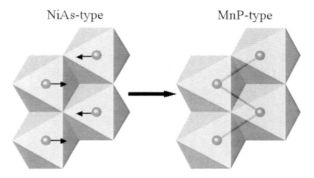

Fig. 12 Displacement of *M* atom positions in MnP-type structure relative to NiAs-type structure, with formation of zigzag chains of *M*–*M* bonds

the geometry changes from exclusively octahedral in ScP, to octahedral and trigonal prismatic in TiP, to trigonal prismatic in the remaining compounds [64]. P–P bonding develops in the form of zigzag chains in the MnP-type compounds (2.6–2.7 Å) and dimers in NiP (2.4 Å) [63, 65]. The *M* atoms are octahedrally coordinated, but *M*–*M* bonding develops as the *M* atoms within regular triangular nets in the NiAs-type structure (VP) approach each other to form zigzag chains in the MnP-type compounds (Fig. 12), which are favoured when the valence electron count lies between 10 and 14 [65].

3.1 Transition-Metal Phosphides, MP

3.1.1 Phosphorus 2p XPS Spectra

The four isostructural MnP-type compounds (CrP, MnP, FeP, CoP) represent an appropriate series for probing how a simple metal substitution (which changes the electron count and the degree of electron transfer between M and P) affects BE shifts, providing information about the various bonding interactions present in the structure. If the Zintl concept is invoked, a simple model for the electronic structure can be proposed in which the M atoms donate sufficient electrons to the anionic framework for each P atom to attain an octet. If $2c-2e^-$ P–P bonds within the zigzag chains are assumed, then a formal charge of P^{1-} obtains (Scheme 1). Such an assignment should, in principle, be distinguishable from the alternative situation where no such bonding is present and only P^{3-} ions are present [65]. Although we can anticipate that full charge transfer to the extreme degree proposed here is clearly unrealistic, the BE shifts on the P atoms can be predicted to *scale* with their charge.

High-resolution phosphorus 2p XPS spectra for MnP-type compounds show a doublet, as seen earlier in a representative spectrum for FeP (Fig. 5), consisting of a lower BE (P $2p_{3/2}$) and a higher BE (P $2p_{1/2}$) peak [58]. The P $2p_{3/2}$ binding energy is lower in this compound than in elemental phosphorus (dashed line, 129.9 eV), clearly indicating the anionic character of the P atom [52, 58]. The same is true not only for other MnP-type compounds but also for all the other first-row transition-metal monophosphides MP [52, 58, 60]. Figure 13 shows a clear trend of lower BE with greater electronegativity difference ($\Delta\chi$) between M and P. That is, on proceeding from CoP to ScP, ionic character in the M–P bond is enhanced, greater charge transfer from M to P occurs, and the P atoms become more negatively charged. By examining BE values (available from the NIST XPS database) for other compounds containing phosphorus in positive (e.g., P_2O_5) oxidation states, as well as for elemental phosphorus itself, we can establish a linear relationship between BE and oxidation state [39]. Interpolation suggests that the P charges in the MnP-type compounds are close to 1– [58]. Although this agrees fortuitously with the Zintl formulation, caution is advised not to interpret the absolute values too literally. The more meaningful conclusion that can be drawn is that the P charges in MnP-type compounds are intermediate between those in elemental phosphorus (more extensive P–P bonding) and ScP (no P–P bonding), and that there is a clear variation of P charges within the MnP-type series. Interestingly, the trend of longer (weaker) P–P bonds within the zigzag chains on going from CrP (2.644 Å) to CoP (2.701 Å),

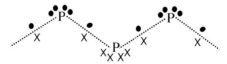

Scheme 1 Lewis structure for homoatomic phosphorus zigzag chain in MnP-type structures containing formal P^{1-} species, if 2c–2e$^-$ bonds are assumed

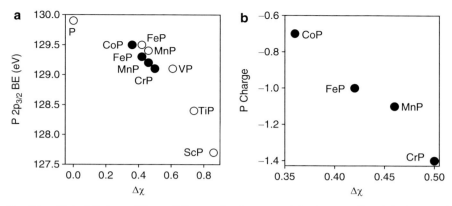

Fig. 13 a Plot of P $2p_{3/2}$ BE vs. difference in electronegativity between P and M in several transition-metal monophosphides MP. This work *filled circle*; [52] *open circle*. **b** Plot of P charge (interpolated from BE values) vs. difference in electronegativity. Reprinted with permission from [58]. Copyright the American Chemical Society

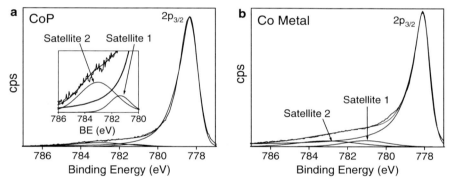

Fig. 14 Co $2p_{3/2}$ spectra for **a** CoP and **b** Co metal. The *inset* in **a** shows the low-intensity satellite structure. Reprinted with permission from [58]. Copyright the American Chemical Society

implying a more negative P charge in the latter, runs counter to the observed trend in BE [66]. Decreased metal-to-P charge transfer (from CrP to CoP) is thus the more dominant effect in influencing the observed BEs [58].

3.1.2 Metal 2p XPS Spectra

The high-resolution metal $2p_{3/2}$ spectra for MnP-type compounds show little, if any, shift in BE relative to the elemental metal, as seen in CoP vs. Co (Fig. 14) [58]. A simplistic interpretation for the absence of a BE shift would be that the metal atoms are neutral in these MnP-type compounds, which is obviously inconsistent with the

1– charge on phosphorus assigned above. A clue in resolving this apparent contradiction is the similar asymmetric lineshape seen in the spectra of both the metal phosphide and the elemental metal. As first proposed by Doniach and Šunjić, this asymmetric line broadening arises from shake-up processes in which valence electrons are promoted just below the Fermi level into a continuum of states in an empty band just above the Fermi level, as occurs in a metallic solid [34]. Moreover, the lack of a significant BE shift is consistent with the delocalization of electrons in conduction states, which allows electrons on the metal atom in MnP-type compounds to experience, on average, greater nuclear screening. The implication, then, is that the metal atoms in MnP-type compounds participate in delocalized bonding within the metal–metal framework, in contrast to the more localized bonding within the P–P zigzag chains. This example illustrates some of the difficulties in interpreting XPS spectra for solids with substantial metallic character, as the BE values alone do not tell the whole story.

How then, can one recover some quantity that scales with the local charge on the metal atoms if their valence electrons are inherently delocalized? Beyond the asymmetric lineshape of the metal $2p_{3/2}$ peak, there is also a distinct satellite structure seen in the spectra for CoP and elemental Co. From reflection electron energy loss spectroscopy (REELS), we have determined that this satellite structure originates from plasmon loss events (instead of a "two-core-hole" final state effect as previously thought [67, 68]) in which exiting photoelectrons lose some of their energy to valence electrons of atoms near the surface of the solid [58]. The intensity of these satellite peaks (relative to the main peak) is weaker in CoP than in elemental Co. This implies that the Co atoms have fewer valence electrons in CoP than in elemental Co, that is, they are definitely cationic, notwithstanding the lack of a BE shift. For the other compounds in the MP ($M = $ Cr, Mn, Fe) series, the satellite structure is probably too weak to be observed, but solid solutions $Co_{1-x}M_xP$ and $CoAs_{1-y}P_y$ do show this feature (*vide infra*) [60, 61].

3.1.3 Valence Band Spectra

The valence band spectra for MnP-type compounds show considerable overlap in energy between metal 3d and phosphorus 3p states, as seen in CoP, for example (Fig. 15) [58]. Nevertheless, these spectra can be fitted with component peaks using a similar protocol as described earlier for $Hf(Si_{0.5}As_{0.5})As$. Peak 1 at high BE represents P 3s states, whereas peaks 2 and 3 represent P $3p_{1/2}$ and $3p_{3/2}$ states, in an intensity ratio of 1:2. Given the octahedral coordination around metal atoms, it seems reasonable to distinguish between metal t_{2g} states (asymmetric peaks 4 and 5 are assigned as $3d_{5/2}$ and $3d_{3/2}$ in an intensity ratio of 3:2) involved in M–P bonding and metal e_g states (single peak 6) involved in M–P antibonding interactions. Peak 7 is important to account for a slight shoulder near the Fermi edge, and is assigned as nonbonding P 3p states, as identified from band structure calculations [58].

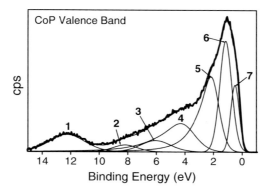

Fig. 15 Fitted valence band spectrum for CoP. Reprinted with permission from [58]. Copyright the American Chemical Society

Table 3 Binding energies (eV) of component peaks in valence band spectra of MP (M = Cr, Mn, Fe, Co)[a]

Peak	Assignment	CrP	MnP	FeP	CoP
1	P 3s	11.8 (3.4) [27.0]	11.7 (3.4) [23.3]	11.9 (3.1) [18.5]	12.3 (2.9) [12.3]
2	P $3p_{1/2}$	8.6 (1.3) [3.6], 5.8 (1.8) [7.3][b]	6.7 (1.9) [8.3]	7.3 (2.4) [3.6]	8.2 (2.0) [3.4]
3	P $3p_{3/2}$	6.8 (1.7) [7.3], 4.3 (2.0) [14.5][b]	4.8 (2.4) [16.6]	5.6 (2.2) [7.2]	6.1 (2.5) [6.8]
4	M $3d_{3/2}$ t_{2g}	3.2 (2.2) [7.8]	3.3 (1.8) [12.0]	3.9 (2.0) [14.4]	4.3 (2.3) [15.7]
5	M $3d_{5/2}$ t_{2g}	2.0 (1.0) [11.7]	2.0 (1.3) [17.8]	2.3 (1.5) [21.5]	2.2 (1.3) [23.5]
6	M $3d_{5/2}$ e_g	1.4 (0.7) [6.1]	1.2 (0.8) [9.7]	1.2 (1.1) [19.0]	1.1 (1.0) [26.9]
7	P 3p	0.6 (1.0) [14.6]	0.5 (0.8) [12.2]	0.5 (0.8) [15.7]	0.5 (0.7) [11.4]

[a]FWHM values (eV) are in parentheses and relative peak areas (%) are in square brackets
[b]A second set of higher binding energy P $3p_{1/2}$ and P $3p_{3/2}$ peaks is present in the spectrum of CrP

Extracting the electron populations by fitting component peaks to the valence band spectra (Table 3) gives charges of approximately $M^{0.7+}$ and $P^{0.7-}$ for all four MnP-type compounds, consistent with the results from core-level spectra. Although the fitting procedure is certainly at its limits of applicability for these spectra, it is interesting that the trends in electron populations do agree with observed changes in bond lengths within this series of compounds. On going from CrP to CoP, the metal t_{2g} states (involved in metal–metal bonding) are increasingly occupied, consistent with the shortening of metal–metal distances (from 2.782 to 2.678 Å), whereas the phosphorus 3p states (involved in P–P bonding) are increasingly depleted, consistent with the lengthening of P–P distances from 2.644 to 2.701 Å [66].

3.2 Mixed-Metal Phosphides, $M_{1-x}M'_xP$

3.2.1 The Charge Potential Model

Even though the effect is small, the discrimination seen in the phosphorus BE values for different members of the MnP-type compounds discussed above suggests that they provide a useful measure of the bonding character (in terms of the degree of electron transfer). To explore this phenomenon further, mixed-metal phosphides $M_{1-x}M'_xP$, where M is designated to be more electronegative than M', can be examined. To a first approximation, the shifts in BE for the phosphorus atoms may be expected to reflect their average environment assuming a random distribution of the two different metal atoms M and M' around them. One of the earliest attempts to quantify BE shifts was formulated by Siegbahn himself, through the *charge potential model* [4]:

$$\Delta E_i = E_i - E_i^0 = k\Delta q_i + \Delta \sum_{j \neq i} q_i / r_{ij} \qquad (6)$$

Here, a photoelectron emanating from a specified core level in an atom i acquires a BE (E_i) that is shifted relative to a reference energy E_i^0 as a result of: (i) intraatomic effects, reflecting the change in charge (Δq_i) around the atom (multiplied by a constant k that scales with interactions between valence and core electrons) and (ii) interatomic effects (or Madelung potential), reflecting the change in the chemical environment ($\Delta \sum_{j \neq i} q_j / r_{ij}$), as defined by the neighbouring atoms each of charge q_j at some distance r_{ij} [4, 5, 69, 70]. Normally, intraatomic effects dominate over interatomic effects, which is why BE often scales simply with atomic charge, especially when the shifts are large. However, interatomic effects may account for more subtle shifts in BE. The immediate coordination environment (nearest neighbours) constitutes the largest contribution to the interatomic term, but because the $1/r_{ij}$ dependence allows interactions at longer range, *next*-nearest neighbours may also exert an effect [5].

It should be emphasized that the charge-potential model assumes that only ground state effects are important in influencing BE shifts [5]. It neglects final state effects in which outer electrons relax towards the nucleus to compensate for the formation of the core hole after photoionization (intraatomic relaxation), which tends to lower the BE apparently observed. Moreover, electron density on neighbouring atoms may also undergo redistribution (interatomic relaxation), which tends to have a greater effect on the BE than intraatomic relaxation [5]. These final state effects modify BE shifts through additional terms, $-(\Delta E_i^{IA1} + \Delta E_i^{IA2})$, that represent intraatomic (IA1) and interatomic (IA2) relaxation, respectively [5]. Applying the charge potential model to interpret BE shifts must therefore be validated only after carefully evaluating these final state effects.

One way to decide if a trend in BE values within a given series of compounds is meaningful is to perform complementary experiments such as XANES. The absorption edges measured from XANES also depend on the same factors as in XPS, providing information about atomic charge and coordination environment [3, 71].

However, in contrast to XPS, electrons are promoted into bound or continuum states and still provide partial screening to the excited atom, so that these absorption energies are less sensitive to relaxation effects. Shifts in absorption energies are thus often more easily detected than those in XPS BEs. Moreover, XANES is a bulk-sensitive technique, and avoids the ambiguities associated with XPS in attributing trends to a bulk vs. surface phenomenon.

3.2.2 Phosphorus 2p XPS and K-edge XANES Spectra

Three series of mixed-metal phosphides, $Co_{1-x}Mn_xP$, $Mn_{1-x}V_xP$, and $Co_{1-x}V_xP$ have been examined by XPS and XANES [60]. In general, their P 2p XPS spectra are similar to those for the binary phosphides (Fig. 5), but the component $2p_{3/2}$ and $2p_{1/2}$ peaks are slightly broadened because different local distributions of metal atoms around each P atom lead to a superposition of multiple signals [60]. The P $2p_{3/2}$ BEs are slightly lower for the mixed-metal phosphides relative to the parent binary phosphides. The P K-edge XANES spectra, shown for $Co_{0.80}V_{0.20}P$ vs. CoP in Fig. 16, also exhibit similar energy shifts, confirming that these changes are derived from a bulk rather than a surface phenomenon. The P $2p_{3/2}$ BE and P K-edge absorption energies are plotted against the difference in electronegativity between phosphorus and metal, where a weighted average, $[(1-x)\chi_M + x\chi_{M'}]$, is used for the latter (Fig. 17a, c) [60]. Both plots show the same trends, with the energies for the mixed-metal phosphides being systematically lower than expected relative to the binary phosphides. Applying the charge potential model provides insight to help explain these trends. As M becomes more electropositive within the series of binary phosphides MP, the charge potential model predicts that the P $2p_{3/2}$ BE will be lowered by the enhanced negative charge q_i on P (because

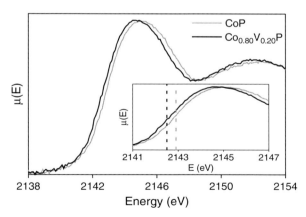

Fig. 16 P K-edge XANES spectra for CoP and $Co_{0.80}V_{0.20}P$. The *inset* highlights the absorption edges, located by their inflection points (*vertical dashed lines*). Reprinted with permission from [60]. Copyright Elsevier

of the greater electron transfer to P within more ionic M–P bonds) but raised by the more positive Madelung potential due to the enhanced positive charges on the surrounding metal atoms. The linear decrease in P $2p_{3/2}$ BE from CoP to TiP confirms that intraatomic effects dominate over interatomic effects. In contrast, interatomic effects do play an important role in influencing the P $2p_{3/2}$ BEs for the mixed-metal phosphides $M_{1-x}M'_xP$. To understand this, it should be recalled that each metal atom in the MnP-type structure is surrounded not only by the six nearest-neighbour P atoms (at 2.2–2.4 Å) but also by four next-nearest-neighbour metal atoms (at 2.6–2.8 Å) forming part of an extended metal–metal bonding network (Fig. 12) [66]. Thus, a given M atom can be surrounded locally not only by other M atoms but also by M' atoms. As a result of the polarization that develops in which the more electropositive M' atoms donate electron density to the more electronegative M atoms (in addition to the P atoms), the M atoms become apparently less positively charged than in the parent binary phosphide MP. The P $2p_{3/2}$ BEs in $M_{1-x}M'_xP$ deviate from the line of best fit for MP because they are additionally modified by the Madelung potential term $\Delta\sum_{j\neq i}\left(q_i/r_{ij}\right)$, which decreases because of the $M' \rightarrow M$ polarization (q_M is less positive at high concentrations of M). The $1/r_{ij}$ dependence in the Madelung term ensures that the charge on the smaller, more electronegative M atoms has a greater influence than the larger, more electropositive M' atoms.

To recover a linear correlation between P $2p_{3/2}$ BEs and difference in electronegativity for the mixed-metal phosphides $M_{1-x}M'_xP$, it is evident that using the weighted average of the electronegativities of M and M' alone is inadequate. A term can be added to represent the charge transfer between these two types of metals, $\chi_M-\chi_{M'}$, which must be multiplied by a scaling factor, $(1-x)/2$, to reflect the proportionate degree of this additional charge transfer [60]:

$$\Delta\chi = \chi_P - [(1-x)\chi_M + x\chi_{M'}] + \frac{(1-x)}{2}(\chi_M - \chi_{M'}) \tag{7}$$

When not only P $2p_{3/2}$ BE values but also P K-edge absorption energies are plotted against this extended expression for the electronegativity difference (Fig. 17c, d), the excellent linear correlations achieved clearly demonstrate the importance of next-nearest neighbour effects in $M_{1-x}M'_xP$. The alternative possibility that these shifts occur because of final state effects instead of ground state effects can also be ruled out. If interatomic adiabatic relaxation of electrons towards the core hole in the final state were significant in affecting the BEs, the energy shifts would actually be more pronounced with greater concentration of electropositive M' atoms [72,73], contrary to observation.

3.2.3 Metal 2p XPS and Mn K-edge XANES Spectra

As in the parent binary phosphides MP, the metal $2p_{3/2}$ XPS spectra for the mixed-metal phosphides $M_{1-x}M'_xP$ exhibit asymmetric lineshapes originating from final state effects [34] involving the metal–metal bonding network. Virtually no changes in BE are observed relative to the binary phosphides MP or the elemental metals

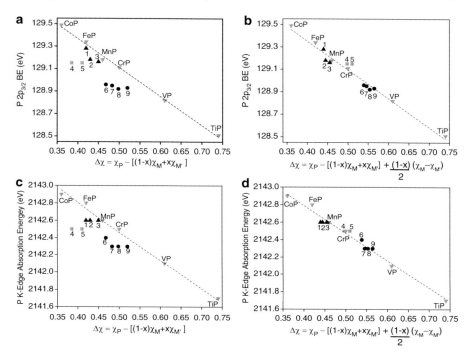

Fig. 17 Plots of **a, b** P $2p_{3/2}$ BE and **c, d** P K-edge absorption energy vs. electronegativity difference for *MP* (*inverted filled triangle*), $Co_{1-x}Mn_xP$ (*filled triangle*) (**1** = $Co_{0.40}Mn_{0.60}P$, **2** = $Co_{0.30}Mn_{0.70}P$, **3** = $Co_{0.10}Mn_{0.90}P$), $Co_{1-x}V_xP$ (*filled square*) (**4** = $Co_{0.90}V_{0.10}P$, **5** = $Co_{0.80}V_{0.20}P$), and $Mn_{1-x}V_xP$ (*filled circle*) (**6** = $Mn_{0.95}V_{0.05}P$, **7** = $Mn_{0.85}V_{0.15}P$, **8** = $Mn_{0.75}V_{0.25}P$, **9** = $Mn_{0.60}V_{0.40}P$), taking into account nearest neighbour contributions only (**a, c**) or with next-nearest contributions included (**b, d**). The electronegativity values are from reference [62]. Reprinted with permission from [60]. Copyright Elsevier

except in the case of the $Co_{1-x}V_xP$ series, where the electronegativity difference between *M* and *M'* is the greatest and a small shift by ~0.2 eV to lower BE is found for the Co $2p_{3/2}$ peak. Although this observation corroborates the $M \rightarrow M'$ charge transfer invoked earlier, the charge potential model cannot be strictly applied here to explain BE shifts because final state effects are now significant. However, we can take advantage of the small satellite peak, found earlier in CoP (Fig. 14) and also seen in all members of the Co-containing mixed-metal series ($Co_{1-x}V_xP$ and $Co_{1-x}Mn_xP$). This satellite peak, which arises from plasmon loss, intensifies with greater concentration of 3d valence electrons on the Co atom and thus reveals information about the Co charge [58]. A plot of the normalized intensity of the plasmon loss peak ($I_{plasmon}/I_{core-line}$) vs. the electronegativity difference ($\chi_M - \chi_{M'}$) indicates that the Co atoms become less positively charged on progressing from CoP to $Co_{1-x}Mn_xP$, and even more so in $Co_{1-x}V_xP$, where the $M \rightarrow M'$ charge transfer is maximized (Fig. 18) [60].

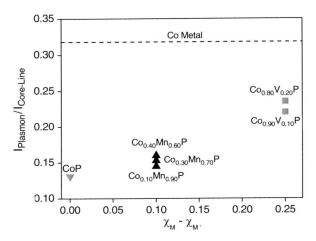

Fig. 18 Normalized intensity of plasmon loss peak in the Co $2p_{3/2}$ spectrum vs. difference in electronegativity for Co-containing phosphides. The *dashed line* indicates the value for Co metal. Reprinted with permission from [60]. Copyright Elsevier

What about the other metal XPS spectra that do not contain observable plasmon loss peaks? In these cases, XANES can also provide information about the occupation of metal-based states. For example, the Mn K-edge XANES spectrum in MnP contains a pre-edge peak, assigned as primarily a quadrupolar transition (1s → 3d) [74], whose intensity decreases with increasing occupancy of the Mn 3d valence states. Because the octahedral coordination of metal atoms is slightly distorted in the MnP-type structure [75], there is also a small dipolar component (1s → 4p) to this peak. For a proper comparison, the Mn K-edge XANES spectra were examined only for those mixed-metal phosphides that are least substituted ($Mn_{0.95}V_{0.05}P$ and $Co_{0.10}Mn_{0.90}P$) (Fig. 19). In $Mn_{0.95}V_{0.05}P$, the pre-edge peak is slightly less intense than in MnP, indicating that the Mn valence states are more occupied and consistent with a V → Mn charge transfer [60]. In contrast, the roles of the metal atoms are reversed in $Co_{0.10}Mn_{0.90}P$, where the pre-edge peak is slightly more intense than in MnP, indicating that the Mn valence states are less occupied and consistent with a Mn → Co charge transfer [60].

3.3 Transition-Metal Arsenides, MAs, and Mixed Arsenide Phosphides, $MAs_{1-y}P_y$

The XPS and XANES investigations of MnP-type binary phosphides MP and mixed-metal phosphides $M_{1-x}M'_xP$ discussed above have revealed some of the factors that influence BEs and absorption energies. In particular, a charge redistribution can take place within the metal–metal bonding network when two different

Fig. 19 Mn K-edge XANES spectra for **a** $Mn_{0.95}V_{0.05}P$ and **b** $Co_{0.10}Mn_{0.90}P$, normalized relative to MnP. Assignments for peaks are: (A) Mn 1s → 3d and 1s → 4p; (B) Mn 1s → P 3p antibonding states; (C) Mn 1s → 4p. Reprinted with permission from [60]. Copyright Elsevier

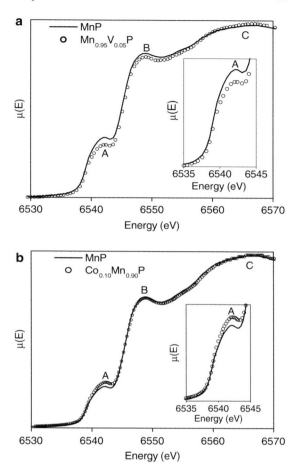

metals are present. In addition to phosphides, the MnP-type structure is adopted by arsenides with some of the first-row transition metals (Cr, Fe, Co) [76]. Because a pnicogen–pnicogen bonding network is also present in this structure, a similar charge transfer may be expected to take place between two different pnicogen atoms, in mixed arsenide phosphides $MAs_{1-y}P_y$. Several binary arsenides MAs (M = V, Cr, Fe, Co) and mixed arsenide phosphides $MAs_{1-y}P_y$ (M = Cr, Fe, Co) have been examined. To interpret the spectra in terms of electronegativity differences in these As-containing compounds, we have chosen a revised value for the electronegativity for As that is intermediate between P and Sb, as discussed in Sect. 4.1.

3.3.1 Phosphorus 2p XPS and K-edge XANES Spectra; Arsenic L-, K-edge XANES Spectra

The P 2p XPS spectra for $M\text{As}_{1-y}\text{P}_y$ resemble those for the parent binary phosphides MP (Fig. 5) but are shifted to lower BE. Unlike the mixed-metal phosphides $M_{1-x}M'_x\text{P}$, however, there is no significant broadening of the component $2p_{3/2}$ and $2p_{1/2}$ peaks [61]. Although the first coordination sphere contains only identical M atoms surrounding a given P photoemission site in a $M\text{As}_{1-y}\text{P}_y$ compound, the second coordination sphere does contain different local distributions of P and As atoms, but they are probably too distant to give rise to distinguishable shifts of the multiple signals that may be expected. The P K-edge XANES spectra also exhibit shifts to lower energy for the mixed arsenide phosphides.

As 3d XPS spectra have also been collected but unfortunately they are not helpful for interpretation because slight reduction of As occurs during the Ar^+-sputtering process and, in the case of CrAs and $\text{CrAs}_{1-y}\text{P}_y$, there is partial overlap with metal 3p signals. Instead, information can be extracted from As L- and K-edge XANES spectra. For example, the absorption edge in the As L_3-edge spectrum for CrAs arises from As $2p \rightarrow 4s$ or $4d$ transitions (Fig. 20). The As K-edge spectra, shown for FeAs and some members of the $\text{FeAs}_{1-y}\text{P}_y$ series (Fig. 21), each contain a strong pre-edge peak (A) and two higher-energy resonances (B and C) [77]. Inspection of the calculated conduction states suggests that peak A is a dipolar As $1s \rightarrow 4p$ transition [61]. Because the 4p states are also involved in bonding, the intensity of this peak reveals information about As charge. Relative to As_2S_3 (containing As in a positive formal oxidation state) [77], this peak is much less intense in MAs and $M\text{As}_{1-y}\text{P}_y$, consistent with the presence of anionic As (with greater occupancy of 4p states). At first glance, peaks B and C could be assigned as As $1s \rightarrow$ As 4p, P 3p, or Fe 4p states, based on the calculated conduction states. However, they are properly assigned as EXAFS peaks, similar to those found in As_2S_3 [77]. This interpretation is supported by the shift of these peaks to higher energy, with greater y in $\text{FeAs}_{1-y}\text{P}_y$ (Fig. 21a), in agreement with the prediction that the energy depends inversely on the scattering path length (shorter average Fe–Pn bond distances) [78].

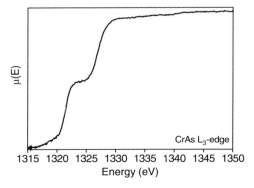

Fig. 20 As L_3-edge XANES spectrum for CrAs, measured in fluorescence mode. Reprinted with permission from [61]. Copyright Elsevier

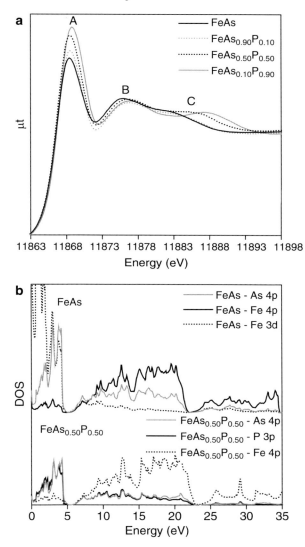

Fig. 21 a Normalized As K-edge XANES spectra for FeAs and some FeAs$_{1-y}$P$_y$ members, measured in transmission mode. **b** Orbital projections of conduction states calculated from FeAs and FeAs$_{0.50}$P$_{0.50}$ (the Fermi edge is at 0 eV). Reprinted with permission from [61]. Copyright Elsevier

The plots of the P 2p$_{3/2}$ BE, P K-edge absorption energy, and As L$_3$-edge absorption energy vs. the electronegativity differences in MAs and MAs$_{1-y}$P$_y$ show interesting trends (Fig. 22). In the binary arsenides MAs, the As L$_3$-edge (Fig. 22c) and K-edge energies (not shown) decrease from CoAs to VAs, similar to the P energies in the binary phosphides MP (Sect. 2.2). The same explanation applies that As atoms are more negatively charged (more negative $k\Delta q_i$ term) as the M–As bond

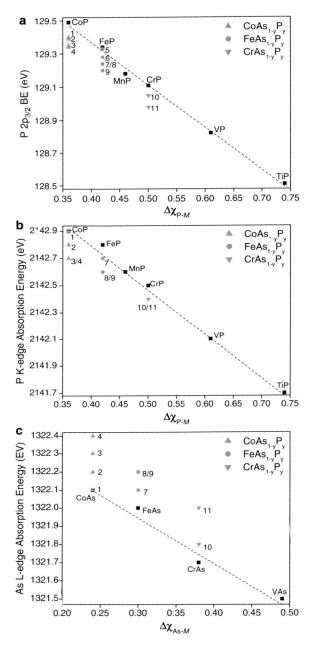

Fig. 22 Plots of **a** P $2p_{3/2}$ BE, **b** P K-edge absorption energy, and **c** As L_3-edge absorption energy vs. electronegativity difference for binary pnictides *M*P or *M*As (*filled square*), and mixed arsenide phosphides CoAs$_{1-y}$P$_y$ (*filled triangle*) (1 = CoAs$_{0.75}$P$_{0.25}$, 2 = CoAs$_{0.50}$P$_{0.50}$, 3 = CoAs$_{0.25}$P$_{0.75}$, 4 = CoAs$_{0.10}$P$_{0.90}$), FeAs$_{1-y}$P$_y$ (*filled circle*) (5 = FeAs$_{0.90}$P$_{0.10}$, 6 = FeAs$_{0.75}$P$_{0.25}$, 7 = FeAs$_{0.50}$P$_{0.50}$, 8 = FeAs$_{0.25}$P$_{0.75}$, 9 = FeAs$_{0.20}$P$_{0.90}$), and CrAs$_{1-y}$P$_y$ (*filled inverted triangle*) (10 = CrAs$_{0.75}$P$_{0.25}$, 11 = CrAs$_{0.10}$P$_{0.90}$). Reprinted with permission from [61]. Copyright Elsevier

becomes more ionic (greater $\Delta\chi = \chi_{As}-\chi_M$). Within a given mixed arsenide phosphide series $MAs_{1-y}P_y$, the P $2p_{3/2}$ BE and P K-edge absorption energies are *lower* than in the parent binary phosphide MP, deviating from the line of best fit with increasing P concentration (y) (Fig. 22a, b). In contrast, the As L_3-edge absorption energies are *higher* than in the parent binary arsenide MAs, deviating from the line of best fit with increasing y (Fig. 22c). Although these shifts are quite small and close to the limit of precision of the measurements, we can attempt to apply the charge potential model to understand these trends as before, attributing them to ground state effects. To do so, relaxation effects must be ruled out. First, the deviations are more pronounced with greater numbers of electronegative P atoms in the second coordination shell, contrary to expectations from relaxation effects [72]. Second, the trends in BE are paralleled by the absorption energies, which are not strongly influenced by relaxation effects. The trends in BEs and absorption energies above indicate that in a given $MAs_{1-y}P_y$ series, the P atoms are more negatively charged than in MP, the As atoms are less negatively charged than in MAs, and that both types of atoms becomes less negatively charged as y increases. With the constraints that the M charge must balance the average Pn charge (which incorporates a concentration factor) and that the P charge is always more negative than the As charge (which is consistent with their relative electronegativities), the charge potential model accounts qualitatively for the observed trends (Fig. 23) [61]. The M atoms in the first coordination shell and the pnicogen atoms in the second coordination shell both modify the Madelung potential operating on the pnicogen centres (P or As), and the BE and absorption energies change accordingly. As y increases, the As atoms become less negatively charged, not only because they give

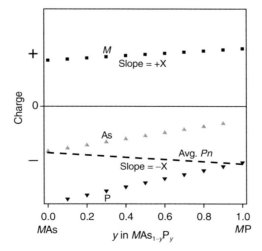

Fig. 23 Dependence of metal and pnicogen charges on level of P substitution (y) in a mixed arsenide phosphide series $MAs_{1-y}P_y$. The average pnicogen charge, $q_{Pn} = yq_P + (1-y)q_{As}$, shown by the *dashed line*, balances the metal charge. Reprinted with permission from [61]. Copyright Elsevier

up electron density to the more electronegative P atoms (via the *Pn–Pn* bonding network) but also because they are inherently less capable than the P atoms to compete for electron density donated by the *M* atoms, which are increasingly positively charged. Moreover, as P atoms gradually substitute for As atoms, the total negative charge is distributed over a greater number of the more electronegative P atoms, such that each P atom, on average, does not need to bear as high a negative charge.

3.3.2 Metal 2p XPS and L-, K-edge XANES Spectra

The metal $2p_{3/2}$ XPS spectra in binary arsenides *M*As exhibit the typical asymmetric lineshape described earlier and the BEs are only slightly lower than in the binary phosphides *M*P, indicating a less positive *M* charge consistent with more covalent *M*–As bonds (Fig. 24a). The mixed arsenide phosphides $MAs_{1-y}P_y$ show intermediate BE shifts, but a better measure of the *M* charge can be extracted from the relative intensity of the satellite peak arising from plasmon loss seen in the $CoAs_{1-y}P_y$ series (Fig. 24b). The plasmon loss intensity, which scales with the Co 3d population, is highest for CoAs and decreases on proceeding to CoP, in agreement with the prediction of increasingly positive metal charge as *y* increases [61].

This trend in metal charge can also be confirmed in the XANES spectra for the $CrAs_{1-y}P_y$ and $FeAs_{1-y}P_y$ series. The Cr K-edge XANES spectra in $CrAs_{1-y}P_y$ resemble the Mn spectrum seen earlier for MnP (Fig. 19), with the pre-edge peak

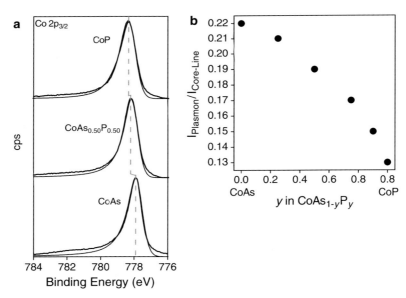

Fig. 24 a Co $2p_{3/2}$ XPS spectra for some members of the $CoAs_{1-y}P_y$ series. **b** Plot of normalized plasmon loss intensity versus *y*. Reprinted with permission from [61]. Copyright Elsevier

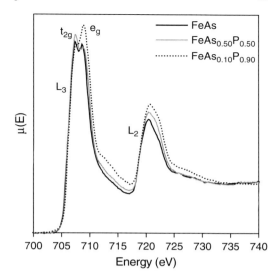

Fig. 25 Normalized total electron yield Fe $L_{2,3}$-edge XANES spectra for FeAs$_{1-y}$P$_y$. The *small shoulders* just above the L_3 and L_2 edges in FeAs$_{0.10}$P$_{0.90}$ probably arise from some surface oxide formed. Reprinted with permission from [61]. Copyright Elsevier

intensifying from CrAs to CrP. The Fe L-edge XANES spectra reveal more intense peaks with more P-rich members in the FeAs$_{1-y}$P$_y$ series (Fig. 25). In particular, there are two resolved L_3 peaks that can be assigned as transitions from the metal 2p to the 3d t_{2g} and e_g states of the octahedrally coordinated Fe atoms. The e_g portion is enhanced in the most P-rich member, FeAs$_{0.10}$P$_{0.90}$, implying that these e_g states are the most depopulated in this series and in agreement with the more positive Fe charge here.

4 Binary and Ternary Skutterudites

The transition-metal monopnictides *MPn* with the MnP-type structure discussed above contain strong *M–M* and weak *Pn–Pn* bonds. Compounds richer in *Pn* can also be examined by XPS, such as the binary skutterudites *MPn*$_3$ (*M* = Co, Rh, Ir; *Pn* = P, As, Sb), which contain strong *Pn–Pn* bonds but no *M–M* bonds [79, 80]. The cubic crystal structure consists of a network of corner-sharing *M*-centred octahedra, which are tilted to form nearly square *Pn*$_4$ rings creating large dodecahedral voids [81]. These voids can be filled with rare-earth atoms to form ternary variants *REM*$_4$*Pn*$_{12}$ (*RE* = rare earth; *M* = Fe, Ru, Os; *Pn* = P, As, Sb) (Fig. 26) [81, 82], the antimonides being of interest as thermoelectric materials [83].

For the binary skutterudites *MPn*$_3$, the electronic structure can be derived from a simple electron-counting scheme by applying the Zintl concept [81]. For example, in CoP$_3$, the more electropositive Co atoms are assigned as low-spin Co^{3+} ($t_{2g}^6 e_g^0$), accounting for its octahedral coordination and the observed diamagnetism [84], and the more electronegative P atoms accept the electrons transferred from the Co atoms

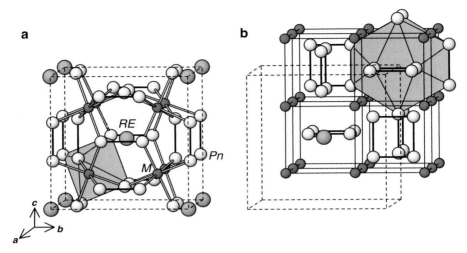

Fig. 26 Skutterudite-type structure in terms of **a** framework of M-centred octahedra and **b** cubic arrangement of M atoms with Pn_4 rings and dodecahedral cages filled with RE atoms in the ternary variants. Reprinted with permission from [110]. Copyright the American Chemical Society

and complete their octets to become P^{1-}, consistent with the presence of P_4 rings with each P atom engaged in two 2c–2e$^-$ P–P bonds. When this model is extended to a band structure, the Co 3d, 4s, and 4p orbitals interact with the P 3s and 3p orbitals to form filled Co–P bonding and Co-based t_{2g} levels at lower energy, and empty Co-based e_g and Co–P antibonding levels at higher energy [85]. These states are superimposed by bands of filled P–P bonding and empty P–P antibonding levels, originating from the P_4 rings. This band structure can be extrapolated to the binary arsenide and antimonide skutterudites. For CoP_3, $CoAs_3$, and $CoSb_3$, most experimental and theoretical studies indicate the existence of a very narrow or zero band gap [56, 86–90].

For the ternary filled skutterudites REM_4Pn_{12}, such as $LaFe_4P_{12}$, the RE atoms can be considered to donate all their valence electrons to the rest of the framework, yielding $La^{3+}[Fe_4P_{12}]^{3-}$. If a P^{1-} charge is assumed in analogy with the binary skutterudites, then the electron count in the $[Fe_4P_{12}]^{3-}$ framework is deficient relative to neutral CoP_3, implying the presence of a hole in the valence band [81, 85]. This hole could reside on the Fe atoms, resulting in mixed valency (three Fe^{2+} ($t_{2g}^6 e_g^0$) and one Fe^{3+} ($t_{2g}^5 e_g^0$) per formula unit), or perhaps within the P–P nonbonding orbitals located close to the Fermi level [81, 85]. Most REM_4Pn_{12} members are indeed found to be hole-doped metals but a few are small band gap semiconductors, such as $CeFe_4P_{12}$ [81, 83, 85, 91–103]. Early arguments suggested a tetravalent state for cerium, which would render the $[Fe_4P_{12}]^{4-}$ framework isoelectronic to binary CoP_3 [82]. Some of these theoretical conclusions conflict with experimental observations. For example, various measurements suggest that $RE\,Fe_4Pn_{12}$ compounds contain exclusively low-spin Fe^{2+}, and that $CeFe_4Pn_{12}$ compounds contain trivalent Ce^{3+} [92, 97, 104–108].

Bonding and Electronic Structure of Phosphides

An attractive feature of applying XPS to study these skutterudites is that the valence states of all atoms can be accessed during the same experiment. As in the study of the MnP-type compounds, these types of investigations also provide insight into bonding character and its relation to electronegativity differences. This information is obtained by analysing both core-line and valence band XPS spectra.

4.1 CoP_3, $CoAs_3$, and $CoSb_3$

Core-line BEs for the pnicogen atoms in $CoPn_3$ are lower than in the elemental pnicogen, but the shifts become less pronounced for the heavier pnictides (P $2p_{3/2}$ BE of 129.3 eV in CoP_3 vs. 130.0 eV in P; As $3d_{5/2}$ BE of 41.3 eV in $CoAs_3$ vs. 41.7(2) eV in As; Sb $3d_{5/2}$ BE of 528.1 eV in $CoSb_3$ vs. 528.2 eV in Sb) [39, 59]. These observations confirm the presence of anionic Pn atoms in $CoPn_3$, as well as the greater covalent character of Co–Pn bonds as electronegativity differences are reduced. For CoP_3, the P $2p_{3/2}$ BE is similar to that found in the binary MnP-type compounds and corresponds to an approximate charge of 1–, agreeing fortuitously with the simple formulation Co^{3+} $(P^{1-})_3$ [59]. For $CoAs_3$, the As $3d_{5/2}$ BE corresponds to a slightly lower pnicogen charge of 0.6–, reflecting the increased covalency of the Co–As bond. For $CoSb_3$, the ionic picture becomes completely unrealistic; donation of electron density from filled Sb orbitals back to empty Co orbitals enhances shielding of the Co nuclear charge and diminishes shielding of the Sb nuclear charge, so that the Sb $3d_{5/2}$ BE in $CoSb_3$ becomes only slightly less than in elemental Sb [59].

The Co 2p spectra for $CoPn_3$ contain $2p_{3/2}$ and $2p_{1/2}$ signals with asymmetric tails as typically observed in metallic compounds [34, 58]. Consistent with the greater screening of the Co nuclear charge with more covalent Co–Pn character, the Co $2p_{3/2}$ BE decreases from 778.5 eV in CoP_3, to 778.3 eV in $CoAs_3$, to 778.1 eV in $CoSb_3$ [61]. If the BE shifts are assumed to be proportional to the degree of covalency in the Co–Pn bond, then this trend implies that the electronegativity of As is intermediate between that of P and Sb [61]. Interestingly, the relative electronegativities of P and As are inconsistent on different electronegativity scales (cf., Allred–Rochow, $\chi_P = 2.06$, $\chi_{As} = 2.20$ vs. Pauling, $\chi_P = 2.2$, $\chi_{As} = 2.1$) [38, 62]. In the electronegativity correlations discussed earlier in the context of MnP-type compounds, we have preferred to use Allred–Rochow values because they are derived from atomic properties and are thus more closely linked to binding and absorption energies [62]. If Allred–Rochow electronegativities are accepted for Co, P, and Sb, and the observed BE shifts are assumed to scale linearly with the electronegativity difference $\Delta\chi = \chi_{Pn} - \chi_M$, then a revised value of $\chi = 1.94$ is obtained for As [61]. Because the Co $2p_{3/2}$ BE in $CoPn_3$ is not much greater than in Co metal (778.1 eV), a better way to gauge the relative charge of the Co atoms is to inspect the intensity of the satellite peak as seen earlier in CoP and CoAs (Sects. 3.1.2 and 3.3.2). If the Co charge is inferred from the pnicogen core-line BEs, they correlate well with the normalized satellite intensity (Fig. 27). The intensity difference between the satellite

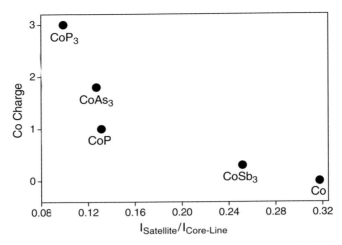

Fig. 27 Correlation between Co charge and intensity of the satellite peak in Co 2p XPS spectra

and core line peaks in CoP$_3$ has been verified by Diplas et al., who propose an alternative explanation in which a two-core-hole process, rather than plasmon loss, is responsible for the Co 2p satellite peak [109]. Notwithstanding the difference in interpretation, the variation in intensity of this satellite peak is clearly related to the Co charge.

4.2 REFe$_4$P$_{12}$ and REFe$_4$Sb$_{12}$

The core-line XPS spectra for LaFe$_4$P$_{12}$ and CeFe$_4$P$_{12}$ reveal P 2p$_{3/2}$ BEs of 129.3 eV, identical to that found in CoP$_3$, and one set of Fe 2p signals with similar asymmetrical lineshapes for the 2p$_{1/2}$ and 2p$_{3/2}$ components, supporting the occurrence of only one Fe valence state [59, 110]. The Fe 2p$_{3/2}$ BE of 707.2 eV is close to that in FeS$_2$ (707.0 eV), which contains Fe^{2+}, and lower than that in K$_3$Fe(CN)$_6$ (709.6 eV), which contains Fe^{3+} [39, 111]. A low-spin state for Fe^{2+} is supported by the observation that the spectra are not broadened by multiplet splitting, which would otherwise occur for high-spin Fe^{2+} [27, 29]. Unlike the P and Fe spectra, the BEs in *RE* 3d spectra generally do not show significant shifts relative to the elemental metal or other compounds (Fig. 28) because the 4f electrons provide poor screening of the nuclear charge. Instead, the lineshapes are characteristic of the valence state [112]. The lineshape in LaFe$_4$P$_{12}$ is similar to that in other La-containing compounds such as LaP, not surprisingly because La is invariably trivalent [110]. However, the lineshape in CeFe$_4$P$_{12}$ more closely resembles that in CeF$_3$ than in CeF$_4$, confirming the presence of exclusively trivalent Ce and ruling out tetravalent Ce, as had been proposed in other studies [110]. These *RE* 3d

Bonding and Electronic Structure of Phosphides 83

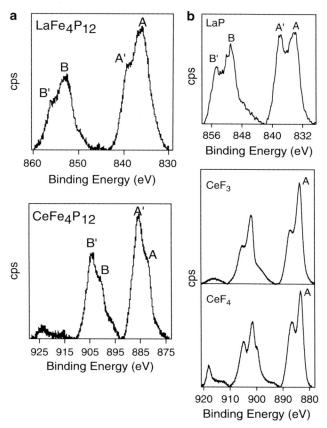

Fig. 28 Comparison of *RE* 3d XPS spectra for **a** LaFe$_4$P$_{12}$ and CeFe$_4$P$_{12}$ with **b** LaP, CeF$_3$, and CeF$_4$. The 3d$_{5/2}$ (*A*) and 3d$_{3/2}$ core lines (*B*), and satellite peaks (*A'*, *B'*) are marked. Reprinted with permission from [110]. Copyright the American Chemical Society

spectra also reveal distinct satellite peaks that can be attributed to ligand-to-metal charge-transfer shake-up processes [32, 33, 110].

The valence band spectra for LaFe$_4$P$_{12}$ and CeFe$_4$P$_{12}$ (Fig. 29) consist of three distinct regions, which can be fitted with the aid of band structure calculations to component atomic orbitals through the procedure described in Sect. 2.4. The assignments are listed in Table 4. The two broad regions, from 15 to 8 eV and from 8 eV to the Fermi edge, represent P 3s and 3p states, respectively. Each of these regions can be fitted by two sets of component peaks, which can be interpreted as separate σ (A) and π (B) components of the P–P bonds within the P$_4$ ring [110]. The third region is a narrow intense band superimposed on the second region from 3 eV to the Fermi edge, and can be fitted by Fe 3d$_{5/2}$ and 3d$_{3/2}$ asymmetric component peaks in an intensity ratio of 3:2. Importantly, the spectrum for CeFe$_4$P$_{12}$ contains an additional component manifested by increased intensity near 2.5 eV, clearly

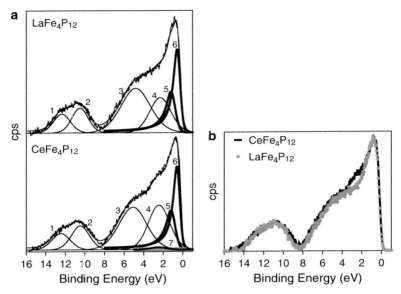

Fig. 29 a Fitted valence band spectra for LaFe$_4$P$_{12}$ and CeFe$_4$P$_{12}$, with peak assignments listed in Table 4. **b** Overlaid spectra highlighting the presence of a Ce 4f component in CeFe$_4$P$_{12}$. Reprinted with permission from [110]. Copyright the American Chemical Society

Table 4 Electron populations in LaFe$_4$P$_{12}$ and CeFe$_4$P$_{12}$

Peak	Assignment	LaFe$_4$P$_{12}$	CeFe$_4$P$_{12}$
1	P 3s A	5.7	5.1
2	P 3s B	7.0	7.5
3	P 3p$_{3/2}$ A	36.5	32.3
4	P 3p$_{3/2}$ B	20.7	26.1
5, 6	Fe 3d$_{3/2}$, 3d$_{5/2}$	25.1	23.6
7	Ce 4f$_{7/2}$		1.4
Total no. of e$^-$		95.0	96.0
Charge per atom		+3 (La), +1.7 (Fe), −0.8 (P)	+2.6 (Ce), +2.1 (Fe), −0.9 (P)

evident when the spectra are overlaid (Fig. 29b). This feature can be interpreted as a Ce 4f$_{7/2}$ peak (filled region), arising from a 4f^1 state located within the valence band and providing direct evidence for the presence of trivalent cerium. Support for this assignment comes from the observation of signals at similar energies in the XPS valence band spectrum for CeF$_3$ and the resonant photoemission spectrum for CeFe$_4$P$_{12}$, as well as the presence of filled 4f states in the calculated valence band of CeFe$_4$P$_{12}$ [92, 108, 110].

Integration of these component peaks, with appropriate corrections applied for different photoionization cross-sections and inelastic mean free paths, gives the electron populations listed in Table 4. The atomic charges obtained are consistent with the results from the core-line spectra and lend further support to the assignment of trivalent RE, low-spin Fe^{2+}, and anionic P^{1-} species.

On proceeding to the antimonides REFe$_4$Sb$_{12}$, greater covalent character can be anticipated to reduce the core-line BE shifts even more. The Sb 3d$_{5/2}$ BEs in LaFe$_4$Sb$_{12}$ and CeFe$_4$Sb$_{12}$ (both 528.0 eV) are similar to that in CoSb$_3$ (528.1 eV) and barely lower than in elemental Sb (528.2 eV) [59]. The Fe 2p$_{3/2}$ BEs are 706.7 eV for LaFe$_4$Sb$_{12}$ and 706.8 eV for CeFe$_4$Sb$_{12}$, which are lower than in the phosphide analogues (707.2 eV) and essentially the same as in Fe metal (706.8 eV) [59]. These small, nearly negligible BE shifts again reflect the significant covalent bonding character in these antimonides relative to the phosphides. Because of this greater covalency and thus greater possibility for interactions between RE and Sb orbitals, there are more possibilities for relaxation mechanisms that complicate the appearance of the RE 3d spectra. Even more complex satellite structures are seen, arising from both ligand-to-metal charge-transfer shake-up processes and two-core-hole processes, but the essential point is that trivalent RE species are confirmed by inspection of the lineshapes of the main core lines [59].

The valence band spectra for LaFe$_4$Sb$_{12}$ and CeFe$_4$Sb$_{12}$ resemble those of the phosphides, but they do not spread over as wide an energy range (12 eV instead of 16 eV) and there is greater mixing of the pnicogen 5p and Fe 3d states in the upper region of the bands (Fig. 30). Although not as prominent as in the phosphides, a component near 2 eV arising from the Ce 4f^1 state is apparent in CeFe$_4$Sb$_{12}$. Integration of these component peaks to obtain electron populations leads to calculated atomic charges (+3 (La), +2.2 (Fe), −1.0 (Sb) for LaFe$_4$Sb$_{12}$ and +2.7 (Ce), +2.4 (Fe), −1.0 (Sb) for CeFe$_4$Sb$_{12}$) in good agreement with expectations from the Zintl formulation [59]. This analysis demonstrates that when core-line spectra do

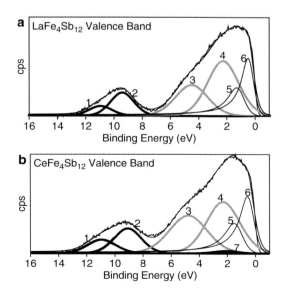

Fig. 30 Fitted valence band spectra for **a** LaFe$_4$Sb$_{12}$ and **b** CeFe$_4$Sb$_{12}$. The peak assignments are analogous to those in Table 4, except that Sb 5s and 5p states account for peaks 1–4. Reprinted with permission from [59]. Copyright the American Physical Society

not reveal significant BE shifts because the bonding is highly covalent, judicious interpretation of the valence band spectra can help recover information about atomic charges.

5 ARXPS Study of Hf($Si_{0.5}As_{0.5}$)As

Up to this point, we have focused on the use of XPS to extract static information about the electronic structure of a solid, taking care to ensure that the measurements are representative of the bulk. In fact, the surface sensitivity of XPS is commonly exploited to determine the surface composition of a solid [1,2]. By taking advantage of the dependence of the photoelectron intensity with the take-off angle (between the surface and the detector), the technique of angle-resolved XPS (ARXPS) allows the surface to be monitored as it undergoes chemical changes [1, 2]. In this way, dynamic information about surface structure can be obtained, which is important for understanding industrially important processes such as corrosion. As an illustration of ARXPS, a kinetic study of the oxidation of Hf($Si_{0.5}As_{0.5}$)As (see Sect. 2.4) is described here. When exposed to air at room temperature, Hf($Si_{0.5}As_{0.5}$)As oxidizes to form hafnium silicates $(HfO_2)_x(SiO_2)_{1-x}$ [113], which have been proposed as high dielectric materials forming a component in semiconductor devices [114,115]. Because the diffusion of foreign elements into other parts of such devices can have a deleterious effect on their performance [116], such studies are often the only means of indirectly probing the mobility of atoms.

The surfaces of large plate-shaped single crystals of Hf($Si_{0.5}As_{0.5}$)As were cleaned by Ar^+ sputtering and then deliberately exposed to the atmosphere over periods of up to 60 min [113]. The Hf 4f, Si 4p, and As 3d spectra were collected for the clean and oxidized surfaces at different take-off angles (Fig. 31). Small amounts of Hf and Si suboxides are already present in the as-cleaned surfaces, presumably by reaction with residual gases in the vacuum chamber of the spectrometer. After oxidation, however, signals originating from several newly formed oxides emerge, while the signals from the substrate diminish substantially. Through comparison with reference standards and literature data [114,117], these new signals were found to correspond to a Si-rich $(HfO_2)_x(SiO_2)_{1-x}$ phase, Hf and Si suboxides, SiO_2, and As_2O_3. Because the surface sensitivity is enhanced at lower take-off angles, the surface composition can be inferred by tracking how the relative intensities of the component peaks change with lower take-off angle. In the Hf 4f spectra, the $(HfO_2)_x(SiO_2)_{1-x}$ peaks are enhanced while the Hf suboxide and Hf($Si_{0.5}As_{0.5}$)As peaks are diminished (the latter even more so). In the Si 2p spectra, the SiO_2 peaks are enhanced while all other peaks are diminished. In the As 3d spectra, the As_2O_3 peaks are enhanced while the Hf($Si_{0.5}As_{0.5}$)As peaks are diminished. As expected, the picture is one where the oxide layers lie on top of the Hf($Si_{0.5}As_{0.5}$)As substrate. For further discrimination among these oxide layers, the O 1s spectra can be inspected (Fig. 32).

Bonding and Electronic Structure of Phosphides

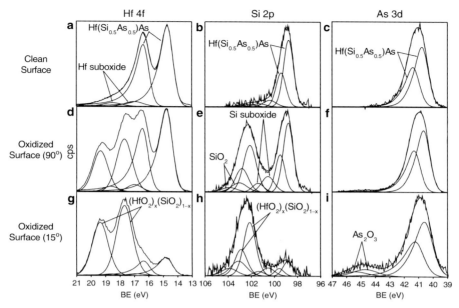

Fig. 31 Core-line XPS spectra for clean and oxidized Hf(Si$_{0.5}$As$_{0.5}$)As at take-off angles of 90° or 15°. Reprinted with permission from [113]. Copyright Wiley

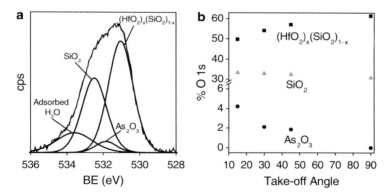

Fig. 32 a O 1s spectrum for oxidized Hf(Si$_{0.5}$As$_{0.5}$)As at a take-off angle of 15°. **b** Plot of O 1s peak intensities vs. take-off angle for different surface components (excluding adsorbed H$_2$O). Reprinted with permission from [113]. Copyright Wiley

The plot of the relative intensities of these signals suggests more SiO$_2$ and As$_2$O$_3$ and less (HfO$_2$)$_x$(SiO$_2$)$_{1-x}$ in the top surface layers. The signals from the Hf and Si suboxides are too weak to be analysed in the same way, but it is reasonable to assume that they are present just at the interface with the substrate. Combining these conclusions leads to the scheme of the surface shown in Fig. 33a [113]. This is a simplistic depiction, since some overlap of the different layers is likely. The formation

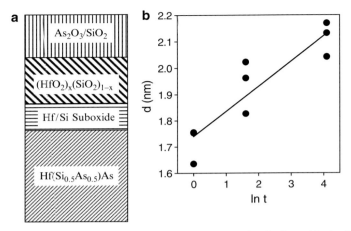

Fig. 33 a Surface composition of oxidized Hf(Si$_{0.5}$As$_{0.5}$)As. **b** Logarithmic kinetics of Hf(Si$_{0.5}$As$_{0.5}$)As oxidation for three different samples. Reprinted with permission from [113]. Copyright Wiley

of As$_2$O$_3$ and SiO$_2$ at the topmost part of the surface and the Si-rich composition of the (HfO$_2$)$_x$(SiO$_2$)$_{1-x}$ layer imply that Si and As are more mobile than Hf atoms through the oxide film. The oxidation of Hf(Si$_{0.5}$As$_{0.5}$)As can be envisioned to proceed by diffusion of Si and As atoms through the intervening (HfO$_2$)$_x$(SiO$_2$)$_{1-x}$ layer via vacancies. The overall thickness of the oxide film, whose composition can be approximated as HfSiO$_4$, can be calculated through the Strohmeier equation from the intensities I in the Hf 4f spectra, the inelastic mean free paths λ, and the volume density of atoms N:

$$d = \lambda_o \sin\Theta \ln\left(\frac{N_m \lambda_m I_o}{N_o \lambda_o I_m} + 1\right) \tag{8}$$

where m refers to the Hf(Si$_{0.5}$As$_{0.5}$)As layer and o to the oxide layer [118]. The logarithmic dependence of the thickness on the time of oxidation (Fig. 33b) confirms that the oxide layer acts as a passive layer while the cations diffuse through it.

6 Conclusion

The studies presented here on several transition-metal phosphides, arsenides, and antimonides illustrate the utility of XPS for obtaining information about bond character, atomic charges, and band structure. In these highly covalent compounds, the BE shifts for the pnicogen atoms can be correlated with the difference in electronegativity between the metal and pnicogen, but these shifts tend to be small and increasingly so on proceeding to the antimonides. Our results suggest that if $\Delta\chi$ is less than 0.2, then the *M–Pn* bond is insufficiently polar to result in much of a BE shift.

However, with the improved resolution of modern XPS instruments, BE shifts as small as $\sim 0.1\,\text{eV}$ can be detected and may be significant. These shifts can be interpreted, to a first approximation, by changes in the atomic charge (an intraatomic effect), but to account for more subtle differences, as seen in the mixed-metal phosphides $M_{1-x}M'_x\text{P}$ and mixed arsenide phosphides $M\text{As}_{1-y}\text{P}_y$, the role of next-nearest neighbours cannot be neglected. These interatomic effects, as incorporated into the charge potential model, help explain the unusual trends in BE observed in these series.

The transition-metal and rare-earth core-line XPS spectra show little, if any, BE shifts at all. Nevertheless, information about atomic charge and valence states can be extracted by examining other features in the spectra. The plasmon loss satellite intensity found in the spectra of Co-containing compounds provides a particularly useful handle on the Co charge. The lineshapes of RE spectra are characteristic of their valence state, as seen in the distinction between trivalent and tetravalent cerium in $\text{CeFe}_4\text{Pn}_{12}$ compounds.

In principle, valence band XPS spectra reveal all the electronic states involved in bonding, and are one of the few ways of extracting an experimental band structure. In practice, however, their analysis has been limited to a qualitative comparison with the calculated density of states. When appropriate correction factors are applied, it is possible to fit these valence band spectra to component peaks that represent the atomic orbital contributions, in analogy to the projected density of states. This type of fitting procedure requires an appreciation of the restraints that must be applied to limit the number of component peaks, their breadth and splitting, and their lineshapes.

XPS has typically been regarded primarily as a surface characterization technique. Indeed, angle-resolved XPS studies can be very informative in revealing the surface structure of solids, as demonstrated for the oxidation of $\text{Hf}(\text{Si}_{0.5}\text{As}_{0.5})\text{As}$. However, with proper sample preparation, the electronic structure of the bulk solid can be obtained. A useful adjunct to XPS is X-ray absorption spectroscopy, which probes the bulk of the solid. If trends in the XPS BEs parallel those in absorption energies, then we can be reasonably confident that they represent the intrinsic properties of the solid. Features in XANES spectra such as pre-edge and absorption edge intensities can also provide qualitative information about the occupation of electronic states.

To extend these investigations, it will be of interest to see if shifts in BE or absorption energies can be discerned in compounds with even less polar covalent bonding, such as those containing more electronegative metals (e.g., Ni, Cu) or less electronegative p-block elements (e.g., Ge, Bi), or in more metal-rich pnictides, such as $M_2\text{P}$ or $M_3\text{P}$. Given that a next-nearest neighbour effect has been identified in both $M_{1-x}M'_x\text{P}$ and $M\text{As}_{1-y}\text{P}_y$, quaternary compounds such as $\text{Mn}_{1-x}\text{Cr}_x\text{As}_{1-y}\text{P}_y$ could also exhibit similar BE shifts. The X-ray spectroscopies presented here provide a valuable addition to the suite of characterization tools available to solid-state chemists, and represent one of a few techniques available to probe directly the electronic structure of solids.

References

1. Briggs D, Grant JT (ed) (2003) Surface analysis by Auger and X-ray photoelectron spectroscopy. SurfaceSpectra, Manchester
2. Briggs D, Seah MP (eds) (1990) Practical surface analysis, vol 1: Auger and X-ray photoelectron spectroscopy, 2nd edn. Wiley, Chichester
3. Wong J, Lytle FW, Messmer RP, Maylotte DH (1984) Phys Rev B 30:5596
4. Siegbahn K, Nordling C, Fahlman A, Nordberg R, Hamrin K, Hedman J, Johansson G, Bergmark T, Karlsson S-E, Lindgren I, Lindberg B (1967) ESCA: atomic, molecular and solid state structure studied by means of electron spectroscopy. Almqvist and Wiksells Boktryckeri, Uppsala
5. Briggs D (2005) Surface analysis of polymers by XPS and static SIMS. Cambridge University Press, Cambridge
6. Brundle CR (1974) J Vac Sci Technol 11:212
7. Belin-Ferré E (2002) Principles and practice. In: Westbrook JH, Fleisher RL (eds) Intermetallic compounds, vol 3. Wiley, New York
8. Watson RE, Perlman ML (1975) X-ray photoelectron spectroscopy, application to metals and alloys. In: Dunitz JD et al. (eds) Photoelectron spectrometry. Structure and bonding, vol 24. Springer, Berlin
9. Wertheim GK (1987) In: Cheetham AK, Day PD (eds) Solid state chemistry: techniques. Oxford University Press, Oxford
10. Watts JF, Wolstenholme J (2003) An introduction to surface analysis by XPS and AES. Wiley, Rexdale
11. Brown GS, Doniach S (1980) In: Winnick H, Doniach S (eds) Synchrotron radiation research. Plenum, New York
12. Brown GS, Calas GA, Waychunas GA, Petiau J (1988) In: Hawthorne FC (ed) Spectroscopic methods in mineralogy and geochemistry. Mineralogical Society of America, Washington
13. Briggs D, Beamson G (1992) Anal Chem 64:1729
14. Kövér L, Novák M, Egri S, Cserny I, Berenyi Z, Tóth J, Varga D, Drube W, Yubero F, Tougaard S, Werner WSM (2006) Surf Interf Anal 38:569
15. Jablonski A, Powell CJ (1994) Phys Rev B 50:4739
16. Vulli M (1981) Surf Interf Anal 3:67
17. Pfau A, Schierbaum KD (1994) Surf Sci 321:71
18. Holgado JP, Alvarez R, Munuera G (2000) Appl Surf Sci 161:301
19. Tougaard S (1990) J Vac Sci Technol A 8:2197
20. Tougaard S (1987) Solid State Commun. 61:547
21. Tanuma S, Powell CJ, Penn DR (2005) Surf Interf Anal 37:1
22. Roberts A (2007) How does the DLD work? Kratos Analytical, Manchester. http://www.kratosanalytical.net/surface. Last accessed 25 Nov 2008
23. Moslemzadeh N, Beamson G, Tsakiropoulos P, Watts JF, Haines SR, Weightman P (2006) J Electron Spectrosc Relat Phenom 152:129
24. Gerson AR, Bredow T (2000) Surf Interf Anal 29:145
25. Koopmans T (1933) Physica 1:104
26. Harris DC, Bertolucci D (1978) Symmetry and spectroscopy: an introduction to vibrational and electronic spectroscopy. Oxford University Press, Oxford
27. Gupta RP, Sen SK (1974) Phys Rev B 10:71
28. Riviere JC (1990) Surface analytical techniques. Oxford University Press, Oxford
29. Grosvenor AP, Kobe BA, Biesinger MC, McIntyre NS (2004) Surf Interf Anal 36:1564
30. Egerton RG, Malac M (2005) J Electron Spectrosc Relat Phenom 143:43
31. Imada S, Jo T (1989) J Phys Soc Jpn 58:402
32. Park K-H, Oh S-J (1993) Phys Rev B 48:14833
33. Sarma DD, Vishnu Kamath P, Rao CNR (1983) Chem Phys 73:71
34. Doniach S, Šunjić M (1970) J Phys C Solid State Phys 3:285
35. Grosvenor AP, Cavell RG, Mar A, Blyth RIR (2007) J Solid State Chem 180:2670

Bonding and Electronic Structure of Phosphides

36. Wang C, Hughbanks T (1995) Inorg Chem 34:5224
37. Tremel W, Hoffmann R (1987) J Am Chem Soc 109:124
38. Pauling L (1960) The nature of the chemical bond, 3rd edn. Cornell University Press, Ithaca, NY
39. Wagner CD, Naumkin AV, Kraut-Vass A, Allison JW, Powell CJ, Rumble Jr JR (2003) NIST X-ray photoelectron spectroscopy database, version 3.4 (web version). National Institute of Standards and Technology, Gaithersburg, MD
40. Fairley N (2003) CasaXPS, version 2.3.9. Casa Software, Teighnmouth, Devon
41. Scofield JH (1976) J Electron Spectrosc Relat Phenom 8:129
42. Yeh JJ, Lindau I (1985) At Data Nucl Data Tables 32:1
43. Sham TK, Rivers ML (2002) Rev Miner Geochem 49:117
44. Green JC, Decleva P (2005) Coord Chem Rev 249:209
45. Wende H (2004) Rep Prog Phys 67:2105
46. Newville M (2004) Fundamentals in XAFS. Consortium for Advanced Radiation Sources, University of Chicago, Chicago
47. Koningsberger DC, Mojet BL, van Dorssen GE, Ramkaer DE (2000) Top Catal 10:143
48. Hähner G (2006) Chem Soc Rev 35:1244
49. Wei S-Q, Sun Z-H, Pan Z-Y, Zhang X-Y, Yan W-S, Zhong W-J (2006) Nucl Sci Tech 17:370
50. Schroeder SLM, Moggridge GD, Rayment T, Lambert RM (1997) J Mol Catal A Chem 119:357
51. Domashevskaya EP, Terekhov VA, Ugai YA, Nefedov VI, Sergushin NP, Firsov MN (1979) J Electron Spectrosc Relat Phenom 16:441
52. Myers CE, Franzen HF, Anderegg JW (1985) Inorg Chem 24:1822
53. Okuda H, Senba S, Sato H, Shimada K, Namatame H, Taniguchi M (1999) J Electron Spectrosc Relat Phenom 101–103:657
54. Shabanova IN, Mitrochin YS, Terebova NS, Nebogatikov NM (2002) Surf Interf Anal 34:606
55. Kanama D, Oyama ST, Otani S, Cox DF (2004) Surf Sci 532:8
56. Anno H, Matsubara K, Caillat T, Fleurial J-P (2000) Phys Rev B 62:10737
57. Kimura A, Suga S, Matsushita T, Daimon H, Kaneko T, Kanomata T (1993) J Phys Soc Jpn 62:1624
58. Grosvenor AP, Wik SD, Cavell RG, Mar A (2005) Inorg Chem 44:8988
59. Grosvenor AP, Cavell RG, Mar A (2006) Phys Rev B 74:125102
60. Grosvenor AP, Cavell RG, Mar A (2007) J Solid State Chem 180:2702
61. Grosvenor AP, Cavell RG, Mar A (2008) J Solid State Chem 181:2549
62. Allred AL, Rochow EG (1958) J Inorg Nucl Chem 5:264
63. Aronsson B, Lundström T, Rundqvist S (1965) Borides, silicides and phosphides. Methuen, London
64. Rundqvist S (1962) Arkiv för Kemi 20:67
65. Tremel W, Hoffmann R, Silvestre J (1986) J Am Chem Soc 108:5174
66. Rundqvist S, Nawapong PC (1965) Acta Chem Scand 19:1006
67. Hüfner S (1995) Photoelectron spectroscopy. Springer, Berlin
68. Raaen S (1986) Solid State Commun 60:991
69. van der Heide PAW (2001) Surf Sci 490:L619
70. van der Heide PAW (2006) J Electron Spectrosc Relat Phenom 151:79
71. de Vries AH, Hozoi L, Broer R (2003) Int J Quantum Chem 91:57
72. Cavell RG, Sodhi RNS (1986) J Electron Spectrosc Relat Phenom 41:25
73. Sodhi RNS, Cavell RG (1986) J Electron Spectrosc Relat Phenom 41:1
74. de Groot FMF, Pizzini S, Fontaine A, Hamalainen K, Kao CC, Hastings JB (1995) Phys Rev B 51:1045
75. Perkins PG, Marwaha AK, Stewart JJP (1981) Theor Chim Acta 59:569
76. Villars P (ed) (2002) Pauling file binaries edition, version 1.0. ASM International, Materials Park, OH
77. Pfeiffer G, Rehr JJ, Sayers DE (1995) Phys Rev B 51:804
78. Fleet ME (2005) Can Miner 43:1811
79. Rundqvist S, Errson N-O (1968) Ark Kemi 30:103

80. Kjekshus A, Rakke T (1974) Acta Chem Scand A 28:99
81. Sales BC (2003) In: Gschneidner KA, Bunzli J-C, Pecharsky VK (eds) Handbook on the physics and chemistry of rare earths. Elsevier, Amsterdam
82. Jeitschko W, Braun D (1977) Acta Crystallogr B 33:3401
83. Nolas GS, Morelli DT, Tritt TM (1999) Annu Rev Mater Sci 29:89
84. Ackermann J, Wold A (1977) J Phys Chem Solids 38:1013
85. Jung D, Whangbo M-H, Alvarez S (1990) Inorg Chem 29:2252
86. Llunell M, Alemany P, Alvarez S, Zhukov VP, Vernes A (1996) Phys Rev B 53:10605
87. Singh DJ, Pickett WE (1994) Phys Rev B 50:11235
88. Lefebvre-Devos I, Lassalle M, Wallart X, Olivier-Fourcade T, Monconduit L, Jumas JC (2001) Phys Rev B 63:125110
89. Sofo JO, Mahan GD (1998) Phys Rev B 58:15620
90. Koga K, Akai K, Oshiro K, Matsuura M (2005) Phys Rev B 71:155119
91. Fornari M, Singh DJ (1999) Phys Rev B 59:9722
92. Nördstrom L, Singh DJ (1996) Phys Rev B 53:1103
93. Danebrock ME, Evers CBH, Jeitschko W (1996) J Phys Chem Solids 57:381
94. Sales BC, Mandrus D, Chakoumakos BC, Keppens V, Thompson JR (1997) Phys Rev B 56:15081
95. Morelli DT, Meisner GP (1995) J Appl Phys 77:3777
96. Singh DJ, Mazin II (1997) Phys Rev B 56:R1650
97. Gajewski DA, Dilley NR, Bauer ED, Freeman EJ, Chau R, Maple MB, Mandrus D, Sales BC, Lacerda AH (1998) J Phys Condens Matter 10:6973
98. Ravot D, Lafont U, Chapon L, Tedenac JC, Mauger A (2001) J Alloys Compd 323–324:389
99. Harima H, Takegahara K (2003) Physica B 328:26
100. Viennois R, Ravot D, Terki F, Hernandez C, Charar S, Haen P, Paschen S, Steglich F (2004) J Magn Magn Mater 272–276: E113
101. Viennois R, Charar S, Ravot D, Haen P, Mauger A, Bentien A, Paschen S, Steglich F (2005) Eur Phys J B 46:257
102. Bauer ED, Chau R, Dilley NR, Maple MB, Mandrus D, Sales BC (2000) J Phys Condens Matter 12:1261
103. Meisner GP, Torikachvili MS, Yang KN, Maple MB, Guertin RP (1985) J Appl Phys 57:3073
104. Grandjean F, Gérard A, Braun DJ, Jeitschko W (1984) J Phys Chem Solids 45:877
105. Long GJ, Hautot D, Grandjean F, Morelli DT, Meisner GP (1999) Phys Rev B 60:7410
106. Xue JS, Antonio MR, White WT, Soderholm L (1994) J Alloys Compd 207–208:161
107. Grandjean F, Long GJ, Cortes R, Morelli DT, Meisner GP (2000) Phys Rev B 62:12569
108. Ishii H, Miyahara T, Takayama Y, Obu K, Shinoda M, Lee C, Shiozawa H, Yuasa S, Matsuda TD, Sugawara H, Sato H (2002) Surf Rev Lett 9:1257
109. Diplas S, Prytz Ø, Karlsen OB, Watts JF, Taftø J (2007) J Phys Condens Matter 19:246216
110. Grosvenor AP, Cavell RG, Mar A (2006) Chem Mater 18:1650
111. Nesbitt HW, Muir IJ (1994) Geochim Cosmochim Acta 58:4667
112. Kaindl G, Wertheim GK, Shmiester G, Sampathkurman EV (1987) Phys Rev Lett 58:606
113. Grosvenor AP, Cavell RG, Mar A (2008) Surf Interf Anal 40:490
114. Barrett N, Renault O, Damlencourt J-F, Martin F (2004) J Appl Phys 96:6362
115. Punchaipetch P, Pant G, Quevedo-Lopez M, Zhang H, El-Bouanani M, Kim MJ, Wallace RM, Gnade BE (2003) Thin Solid Films 425:68
116. Wilk GD, Wallace RM, Anthony JM (2001) J Appl Phys 89:5243
117. Morant C, Galan L, Sanz JM (1990) Surf Interf Anal 16:304
118. Strohmeier BR (1990) Surf Interf Anal 15:51
119. Anderson OK (1975) Phys Rev B 12:3060

Struct Bond (2009) 133: 93–120
DOI:10.1007/430_2008_12
© Springer-Verlag Berlin Heidelberg 2009
Published online: 5 March 2009

Oxo-Centered Triruthenium-Acetate Cluster Complexes Derived from Axial or Bridging Ligand Substitution

Zhong-Ning Chen and Feng-Rong Dai

Abstract The oxo-centered triruthenium-acetate cluster complexes with general formula $[Ru_3(\mu_3\text{-}O)(\mu\text{-}OAc)_6(L)_2L']^{n+}$ (L and L' = axial ligand, $n = 0$, 1, 2) show attractive ligand substitution reactivity, multiple redox behavior and rich mixed-valence chemistry. The axial ligands L and L' are comparatively labile and readily substitutable. Although the $Ru_3(\mu_3\text{-}O)(\mu\text{-}OAc)_6$ cluster core possesses high stability, displacement of one of the bridging acetates has been achieved by using π-delocalized N-hetrocyclic ligands with low π^* energy levels. Ligand substitution not only affords an excellent means of tuning the redox levels of electron transfer processes, but also provides a feasible approach to design ligand-linked triruthenium cluster oligomers with desired properties. This article reviews the recent progress in the ligand substitution chemistry of oxo-centered triruthenium–acetate complexes with parent $Ru_3(\mu_3\text{-}O)(\mu\text{-}OAc)_6$ cores. The syntheses, redox and spectroscopic properties, and mixed valence chemistry of these oxo-centered triruthenium cluster derivatives are summarized to correlate structures with properties.

Keywords: Acetate · Cluster · Electronic interaction · Ligand substitution · Redox · Triruthenium

Contents

1	Introduction	94
2	Triruthenium Complexes from Axial Ligand Substitution	95
	2.1 N-Heterocyclic Ligand-Substituted Triruthenium Complexes..................	96
	2.2 Phosphine Ligand-Substituted Triruthenium Complexes	102

Z.-N. Chen (✉) and F.-R. Dai
State Key Laboratory of Structural Chemistry, Fujian Institute of Research on the Structure of Matter, Chinese Academy of Sciences, Fuzhou, Fujian 350002, China
e-mail: czn@fjirsm.ac.cn

3	Triruthenium Complexes from Bridging Ligand Substitution	108
	3.1 Triruthenium Complexes with Ortho-Metallated Ligands	108
	3.2 Azo–Aromatic Ligand-Substituted Triruthenium Complexes	116
4	Concluding Remarks	118
References		119

Abbreviations

2,2′-bpy	2,2′-Bipyridine
4,4′-bpy	4,4′-Bipyridine
abco	1-Azabicyclo[2,2,2]octane
abcp	2,2′-Azo-bis(5-chloropyrimidine)
abpy	2,2′-Azo-bispyridine
BPA	1,2-Bis(4-pyridyl)ethane
BPE	*Trans*-1,2-bis(4-pyridyl)ethylene
bpyC≡Cbpy	Bis(2,2′-bipridin-5-yl)ethyne
bpyC≡C–C≡Cbpy	Bis(2,2′-bipyridin-5-yl)butadiyne
bpym	2,2′-Bipyrimidine
bpz	2,2′-Bipyrazine
Br$_2$bpy	5,5′-Dibromo-2,2′-bipyridine
CLCT	Cluster-to-ligand charge transfer
cpy	4-Cyanopyridine
dbbpy	4,4′-Dibutyl-2,2′-bipyridine
dmap	4-Dimethylaminopyridine
Dmbpy	4,4′-Dimethyl-2,2′-bipyridine
IC	Intracluster charge transfer
IVCT	Intervalence charge transfer
mbpy$^+$	*N*-Methyl-4,4′-bipyridinium ion
OAc	Acetate
phen	1,10-Phenanthroline
PPh$_3$	Triphenylphosphine
py	Pyridine
pyq	Pyrazino[2,3-f]quinoxaline
pz	Pyrazine

1 Introduction

It has been several decades since oxo-centered triruthenium–carboxylate complexes with triangular cluster frameworks of $Ru_3(\mu_3\text{-O})(\mu\text{-OOCR})_6$ (R = alkyl or aryl) were first isolated [1,2]. In the early 1970s, the first oxo-centered triruthenium complex was structurally characterized by Cotton through X-ray crystal structural determination [3]. Since then, oxo-centered trinuclear ruthenium–carboxylate cluster complexes with general formula $[Ru_3O(OOCR)_6(L)_2L']^{n+}$ (R = aryl or alkyl, L and

$L' =$ axial ligands, $n = 0, 1, 2$) have been extensively investigated regarding their specific electronic, chemical, and physical properties [4]. Particularly, oxo-centered triruthenium cluster complexes with bridging acetates attracted the most attention owing to their synthetic accessibility, multiple redox behavior, intriguing mixed-valence chemistry, and versatile catalytic properties [5–7].

One of the most remarkable features in oxo-centered triruthenium–acetate cluster complexes is ligand substitution reactivity [4–7]. On the one hand, since the axial ligands in triruthenium cluster complexes $[Ru_3O(OAc)_6(L)_2L']^{n+}$ are relatively labile, substitution of them by specifically selected ligands not only affords an available means of controlling their electronic, spectroscopic, and redox properties, but also provides an excellent approach to design ligand-linked triruthenium cluster oligomers with desired properties. One the other hand, the six bridging acetates are relatively stable because of the presence of six six-membered coordination rings in the triruthenium cluster skeleton $Ru_3(\mu_3\text{-}O)(\mu\text{-}OOCR)_6$. It appears that replacement of the bridging acetates through ligand substitution is quite difficult and needs rigorous reaction conditions if possible. In fact, displacement of the bridging acetates in the oxo-centered triruthenium cluster $Ru_3(\mu_3\text{-}O)(\mu\text{-}OAc)_6$ had never been reported in the literature until triruthenium cluster derivatives with *ortho*-metallated $2,2'$-bipyridyl ligands were isolated in 2004 by reaction of methanol-containing precursor $[Ru_3O(OAc)_6(py)_2(CH_3OH)]^+$ with $2,2'$-bipyridyl ligands at ambient temperature [8]. This discovery concerning substitution of one of six bridging acetates in the parent $Ru_3(\mu_3\text{-}O)(\mu\text{-}OAc)_6$ cluster not only opens a significant synthetic route to access oxo-centered triruthenium derivatives with $Ru_3(\mu_3\text{-}O)(\mu\text{-}OAc)_5$ cluster moieties, but also affords another promising approach to control their electronic, optical, and magnetic properties through bridging ligand substitution.

2 Triruthenium Complexes from Axial Ligand Substitution

Oxo-centered triruthenium complexes with different axial ligands can usually be prepared by reactions of solvent-coordinated triruthenium precursors with a wide range of ligands. As the axial solvent molecules are weakly bonded to the ruthenium centers, ligand-substituted triruthenium derivatives are usually accessible at ambient temperature. Depending on the structures and types of desired triruthenium derivatives, the solvent-coordinated triruthenium complexes with different number of solvent molecules and formal oxidation states can be selected as the synthetic precursors, including $[Ru_3O(OAc)_6(CH_3OH)_3]^+$ (**1**), $[Ru_3O(OAc)_6(py)_2(CH_3OH)]^+$ (**2**), $[Ru_3O(OAc)_6(CO)(CH_3OH)_2]$ (**3**), $[Ru_3O(OAc)_6(CO)(py)(H_2O)]$ (**4**), etc. [9, 10]. By judicious selection of axial ligands, a wide range of monomeric, oligomeric, and polymeric oxo-centered triruthenium species with varied nuclearity and topology have been prepared through elaborately designed synthetic approaches [7]. These oxo-centered triruthenium derivatives exhibit the desired electronic, optical, and redox properties, which are tunable by modification of the substituents in the axial ligands.

Of various inorganic and organic ligands, N-heterocyclic and phosphine ligands are mostly utilized for the preparation of ligand-substituted oxo-centered triruthenium-acetate complexes because these neutral N or P donor ligands can afford strong σ-donation as well as π-back bonding when bonded to oxo-centered triruthenium clusters.

2.1 N-Heterocyclic Ligand-Substituted Triruthenium Complexes

N-heterocycles are a class of neutral ligands with strong coordination affinity to many metal ions. Since a number of neutral N-donors ligands are available, a wide range of oxo-centered triruthenium complexes with various N-heterocyclic ligands have been prepared through axial ligand substitution. By judicious selection of the N-heterocyclic type and modification of the substituents with different electronic and steric effects, the electronic, redox, and spectroscopic properties in these oxo-centered triruthenium derivatives are controllable.

2.1.1 Pyridyl Ligand-Substituted Triruthenium Complexes

Using solvent-containing triruthenium species **1** as a synthetic precursor, a series of pyridyl-substituted triruthenium derivatives $[Ru_3O(OAc)_6(py)_2(L)]^+$ (L = 4,4′-bpy **5**, BPE **6**, BPA **7**) were prepared by Meyer et al. [9]. Electrochemical studies showed that these triruthenium complexes exhibit four to five reversible one-electron redox waves in the potential range of $+2.0$ to -2.0 V, suggesting that these complexes can exist in a series of redox states, including -3, $+2$, $+1$, 0, -1, and -2. It is noteworthy that the neighboring redox waves are always separated by ca 1 V, indicating an extensive electronic delocalization in the oxo-centered triruthenium molecules. Chemical or electrochemical oxidation or reduction of $[Ru_3^{III,III,III}]^+$ complexes induced isolation of stable $[Ru_3^{IV,III,III}]^{2+}$ or $[Ru_3^{III,III,II}]^0$ species. These triruthenium species with different oxidation states or electron contents afford intense absorptions in the visible region due to a series of closely spaced electronic transitions within the triruthenium cluster core. Both the multiplet redox behavior and spectral features can be interpreted in terms of a delocalized Ru_3O-based π-molecular orbital system. Because of the presence of strong Ru–Ru interactions through the central μ_3-O atom, there exist a series of delocalized Ru–Ru and Ru–O–Ru levels based on the Ru_3O cluster core.

By summarizing the relationship between redox potentials of a series of $[Ru_3O(OAc)_6(L)_3]^+$ complexes and pK_a of N-heterocyclic ligands L [11], it is demonstrated that the related redox potentials $E_{1/2}$ exhibit a linear energy correlation with the pK_a values according to the equations $E_{1/2}^{(+3/+2)} = 2.24 - 0.023pK_a$; $E_{1/2}^{(+2/+1)} = 1.34 - 0.029pK_a$; $E_{1/2}^{(+1/0)} = 0.36 - 0.039pK_a$, and $E_{1/2}^{(0/-1)} = -0.68 - 0.074pK_a$. As the redox potentials decrease with the pK_a of the N-heterocyclic ligands, with increase of the pK_a by the sequence pyrazine $<$ methylpyrazine $<$ 2,6-dimethylpyrazine $<$ aminopyrazine $<$ 4-acetylpyridine $<$ pyridine $<$ 4-*tert*-butylpyridine $<$ aminopyridine, the higher oxidation states are stabilized mainly via σ-bonding effects, inducing a progressive decrease of the $E_{1/2}$ values along this sequence. Alternately, the triruthenium complexes at lower oxidation state are further stabilized by π-back bonding interactions from the axial π-accepting N-heterocyclic ligands, leading to a higher dependence of $E_{1/2}$ on pK_a for the lower oxidation state species.

By introducing redox-active N-methyl-4,4′-bipyridinium ion (mbpy$^+$) to the oxo-centered triruthenium cores, a series of triruthenium derivatives bearing two or three axially coordinated mbpy$^+$ were prepared by Abe et al. [12, 13]. Electrochemical studies indicated that these mbpy$^+$-containing triruthenium complexes afforded a total of seven to nine reversible or quasi-reversible redox waves in acetonitrile solutions at ambient temperature. Of these redox waves, four or five one-electron redox processes arise from Ru$_3$-based oxidations or reductions involving five or six formal oxidation states, including

$Ru_3^{IV,IV,III}/Ru_3^{IV,III,III}/Ru_3^{III,III,III}/Ru_3^{III,III,II}/Ru_3^{III,II,II}/Ru_3^{II,II,II}$, whereas others are from mbpy$^+$ ligand-centered reductions. Remarkably, observation of two couples of distinctly spaced one-electron waves due to ligand-centered processes mbpy$+$/mbpy$^{\cdot}$ and mbpy$^{\cdot}$/mbpy$^-$ revealed that distinct ligand–ligand electronic interaction is operating through the electronically delocalized oxo-centered triruthenium cluster core.

Through reaction of two equivalents of methanol-containing precursor [Ru$_3$ O(OAc)$_6$(py)$_2$(CH$_3$OH)]$^+$ with L (L = 4,4'-bpy, BPE and BPA), Meyer et al. prepared a series of bipyridine-linked cluster dimers [{Ru$_3$O(OAc)$_6$(py)$_2$}$_2$(μ-L)]$^{2+}$ [9]. Electrochemical studies showed that the distinct redox wave splitting between two Ru$_3$ cluster moieties was not observed. The UV–vis-NIR spectroscopic studies indicated that their absorption features are the same as those of the individual L-containing triruthenium species [Ru$_3$O(OAc)$_6$(py)$_2$(L)]$^+$ except that the molar extinction coefficient in the dimers is doubled. It appears that two Ru$_3$ cluster units in the dimeric species are electronically isolated without cluster–cluster interaction through the bridging 4,4'-bipyridine ligands.

Ito et al. [14] prepared a series of 4,4'-bipyridine linked dimeric triruthenium cluster species [{Ru$_3$O(OAc)$_6$(CO)(L')}$_2$(μ-4,4'-bpy)] (L' = dmap **8**, py **9**, or cpy **10**) by reaction of H$_2$O-containing precursor [Ru$_3$O(OAc)$_6$(CO)(L')(H$_2$O)] with one equivalent of 4,4'-bpy terminally coordinated species [Ru$_3$O(OAc)$_6$ (CO)(L') (4,4'-bpy)]. By modifying the axial ligand L', redox interaction due to two successive $Ru_3^{III,III,II}/Ru_3^{III,II,II}$ processes is tunable. The potential difference $\Delta E_{1/2}$ due to redox splitting of two identical $Ru_3^{III,III,II}/Ru_3^{III,II,II}$ processes is 0.12 V for **8**, 0.08 V for **9**, and 0.05 V for **10**. Obviously, introducing an electron-donating substituent to axial pyridine ligand promotes the cluster-cluster electronic interaction, whereas an electron-withdrawing substituent reduces the intercluster interaction.

8 R = N(CH$_3$)$_2$
9 R = H
10 R = CN

Layer-by-layer Ru$_3$ cluster-based multilayers were fabricated onto preorganized self-assembled monolayer gold electrode surfaces by Abe et al. [15], in which [Ru$_3$(μ$_3$-O)(μ-OAc)$_6$(4,4'-bpy)$_2$(CO)] was utilized as the synthetic precursor. The stepwise connection of oxo-centered triruthenium cluster units onto the gold electrode surface is a feasible approach for construction of Ru$_3$ cluster-based oligomers on a solid surface, in which the bridging ligand 4,4'-bipyridine appears to mediate weak cluster-cluster electronic interaction between the Ru$_3$ cluster centers.

2.1.2 Pyrazine-Substituted Triruthenium Complexes

Pyrazine-Linked Dimers of Ru$_3$ Units

Pyrazine (pz) is one of the most efficient ditopic N-heterocyclic ligands for transmitting magnetic and electronic interactions between various metal centers. Meyer et al. [16, 17] first prepared pyrazine-linked oxo-centered dimer [{Ru$_3$O(OAc)$_6$(py)$_2$}$_2$ (μ-pz)]$^{2+}$ (**11**) by reaction of η^1-pyrazine-coordinated triruthenium species [Ru$_3$O (OAc)$_6$(py)$_2$(pz)]$^+$ with one equivalent of methanol-coordinated triruthenium precursor **2** through formation of the Ru$_3$-pz-Ru$_3$ linkage. Cyclic voltammetry studies showed that distinct redox wave splitting occurs in the dimer **11** compared with that in the corresponding η^1-pyrazine triruthenium species [Ru$_3$O(OAc)$_6$(py)$_2$(pz)]$^+$, arising most likely from cluster-cluster electronic interaction. The extent of intercluster interaction is dependent on axial ligands as well as oxidation states or electron content of the triruthenium cluster units. For the species involving Ru$_3^{IV,IV,III}$/ Ru$_3^{IV,III,III}$ and Ru$_3^{IV,III,III}$/Ru$_3^{III,III,III}$ states, redox wave splitting is irresolvable by cyclic voltammetry. For the species with lower oxidation states, the redox potential difference $\Delta E_{1/2}$ from redox wave splitting is 0.10 V for the Ru$_3^{III,III,III}$/Ru$_3^{III,III,II}$ state and 0.27 V for the Ru$_3^{III,III,II}$/Ru$_3^{III,II,II}$ state. This suggests that cluster–cluster electronic interaction is progressively enhanced with decrease of the oxidation states in the Ru$_3$ cluster units because the extent of cluster–π^*(ligand) mixing will increase as the electron content in the cluster increases. As a result, electronic delocalization is likely the greatest for the 1– mixed-valence dimeric species [Ru$_3^{III,III,II}$-pz-Ru$_3^{III,II,II}$]$^-$. For the intercluster mixed-valence [Ru$_3^{III,III,III}$-pz- Ru$_3^{III,III,II}$]$^+$ species that was stably isolated at ambient temperature, the electronic spectrum includes absorption envelopes characteristic of both Ru$_3^{III,III,III}$ and Ru$_3^{III,III,II}$ species, which are only slightly perturbed in the dimer. The absorption from intervalence charge transfer (IVCT) transition in the intercluster mixed-valence species [Ru$_3^{III,III,III}$-pz-Ru$_3^{III,III,II}$]$^+$, however, is severely overlapped by intense absorptions from intracluster transitions (IC) in the near infrared region.

11

Ito and Kubiak et al. [14, 18–21] described the preparation of pyrazine-linked triruthenium dimers [{Ru$_3$O(OAc)$_6$(CO)(L)}$_2$(μ-pz)] (L = abco **12**, dmap **13**, py **14**, cpy **15**,) by reaction of CO-containing precursor [Ru$_3$O(OAc)$_6$(CO)(L)(pz)] with one equivalent of H$_2$O-coordinated triruthenium complex [Ru$_3$O(OAc)$_6$

(CO)(L)(H$_2$O)]. Redox wave splitting from two identical one-electron reduction processes Ru$_3^{III,III,II}$/Ru$_3^{III,II,II}$ is tunable by modification of the substituent in axial pyridyl ligands. Redox potential difference $\Delta E_{1/2}$ due to cluster–cluster interaction is 0.47 V for **12** (L = abco), 0.44 V for **13** (L = dmap), 0.38 V for **14** (L = py), and 0.25 V for **15** (L = cpy). Obviously, introducing electron-donating substituent to the axially coordinated pyridyl ligands favors electronic delocalization between two Ru$_3$ cluster units, whereas an electron-withdrawing substituent reduces the intercluster interaction. Intercluster mixed-valence species [Ru$_3^{III,III,II}$-pz-Ru$_3^{III,II,II}$]$^-$ exhibits an IVCT band at 12,500 cm^{-1} for **12**, 12,100 cm^{-1} for **13**, 11,800 cm^{-1} for **14**, and 10,800 cm^{-1} for **15**. The ν(CO) vibrational frequency in intercluster mixed-valence species [{Ru$_3^{III,III,II}$-pz-Ru$_3^{III,II,II}$}]$^-$, measured by reflectance IR spectroelectrochemistry, is highly dependent on the axial ligands. The rate constant k_{ET} of intercluster electron transfer in the mixed valence state of **12–15** was estimated by simulating dynamical effects on the ν(CO) band shape with $k_{ET} = 1 \times 10^{12}$ for **12**, 9×10^{11} for **13**, 5×10^{11} for **14**, and 1×10^{11} for **15**. This implies cluster–cluster electron transfer in the intercluster mixed-valence species [{Ru$_3^{III,III,II}$-pz-Ru$_3^{III,II,II}$}]$^-$ being in the infrared vibrational timescale.

12 R = abco
13 R = N(CH$_3$)$_2$
14 R = H
15 R = CN

Pyrazine-Linked Oligomers of Ru$_3$ Units

A series of pyrazine-linked trimers, tetramers, and hexamers have been accessed through elaborately designed synthetic approaches. Meyer et al. [22] first reported the isolation of the pyrazine-linked trimeric species [{Ru$_3$O(OAc)$_6$(py)$_2$} (μ-pz){Ru$_3$O(OAc)$_6$(CO)}(μ-pz){Ru$_3$O(OAc)$_6$(py)$_2$}]$^{2+}$ (**16**) by reaction of η^1-pyrazine-coordinated complex [Ru$_3$O(OAc)$_6$(py)$_2$(pz)]$^+$ with two equivalents of two methanol-coordinated complex [Ru$_3$O(OAc)$_6$(CO)(CH$_3$OH)$_2$] (**3**). Compared with those in the η^1-pyrazine triruthenium complexes [Ru$_3$O(OAc)$_6$ (py)$_2$(pz)]$^+$ (+2.03, +1.04, 0.00, −1.20, and −1.88 V) and [Ru$_3$O(OAc)$_6$(CO) (pz)$_2$] (+1.36, +0.69, −0.66, and −1.30), ten reversible redox waves in the trimeric species **16** were tentatively assigned. The waves at +2.05,

+1.06, +0.04, −1.05, −1.37, and −1.56 V are likely due to two Ru_3 units at each end, whereas the waves at +1.49, +0.72, −0.63, and −1.21 V arise probably from the central Ru_3 unit with axially coordinated CO. Occurrence of several closely spaced one-electron waves at the cathodic side is probably induced by the cluster–cluster interaction because intercluster coupling is always enhanced with increase of the oxidation states or electron contents in the Ru_3 units.

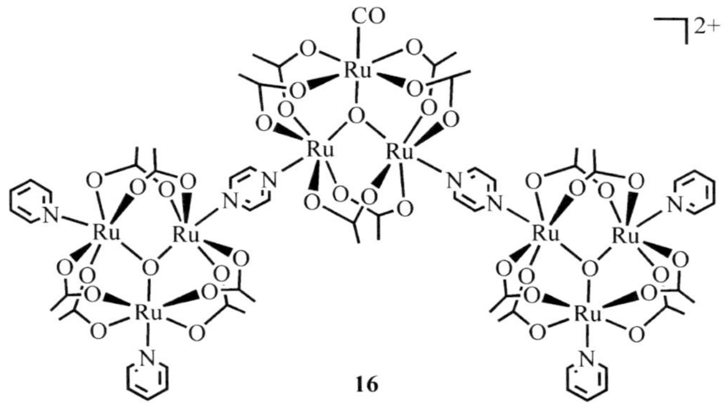

16

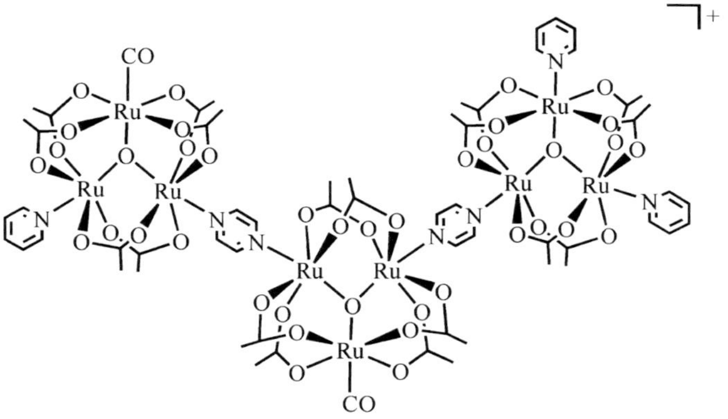

17

With different synthetic routes by judicious selection of various Ru_3 cluster moieties as synthetic precursors, Ito et al. and Toma et al. [10, 23, 24] prepared a series of asymmetric or symmetric pyrazine-linked Ru_3 cluster trimers and tetramers that exhibit multistep and multielectron reversible redox processes. Asymmetric trimeric species $[\{Ru_3O(OAc)_6(CO)(py)\}(\mu\text{-pz})\{Ru_3O(OAc)_6(CO)\}(\mu\text{-pz})\{Ru_3O(OAc)_6(py)_2\}]^+$ (**17**) was prepared through a two-step synthetic approach. The first step involves preparation of methanol-containing asymmetric dimeric species $\{Ru_3O(OAc)_6(CO)(py)\}(\mu - pz)\{Ru_3O(OAc)_6(CO)(CH_3OH)\}$ by reaction of η^1-pyrazine species $Ru_3O(OAc)_6(CO)(py)(pz)$ with $Ru_3O(OAc)_6(CO)(CH_3OH)_2$. The desired trimer **17** was then prepared by incorporating the dimer

$\{Ru_3O(OAc)_6(CO)(py)\}(\mu\text{-}pz)\{Ru_3O(OAc)_6(CO)(CH_3OH\}$ with η^1-pyrazine species $[Ru_3O(OAc)_6(py)_2(pz)]^+$. The trimer **17** exhibits 11 reversible redox waves, in which ten of them are from one-electron processes and one is from a two-electron step composed of two overlapping one-electron redox waves. Each Ru_3 cluster unit affords four reversible one-electron waves involving $Ru_3^{IV,III,III}$ to $Ru_3^{II,II,II}$ states. The possible cluster–cluster interaction through the bridging pyrazine makes the potential difference for each Ru_3 cluster units more pronounced in the cathodic potential region. As found in pyrazine-linked Ru_3 cluster dimers, intercluster electronic interaction is always enhanced with increase of the oxidation states or electron contents in the individual Ru_3 cluster unit.

18

Ito et al. [23] described the preparation of asymmetric tetramer $[\{Ru_3O(OAc)_6(py)(CO)\}(\mu\text{-}pz)\{Ru_3O(OAc)_6(CO)\}(\mu\text{-}pz)\{Ru_3O(OAc)_6(py)\}(\mu\text{-}pz)\{Ru_3O(OAc)_6(dmap)_2\}]^{2+}$ (**18**) by incorporating asymmetric neutral dimer $\{Ru_3O(OAc)_6(py)(CO)\}(\mu\text{-}pz)\{Ru_3O(OAc)_6(CO)(solvent)\}$ with 2+ asymmetric dimer $[\{Ru_3O(OAc)_6(py)(pz)\}(\mu\text{-}pz)\{Ru_3(Ru_3O(OAc)_6(dmap)_2\}]^{2+}$ through displacing the axial solvent molecule in the former by η^1-pyrazine in the latter. This Ru_3 cluster tetramer exhibits 14 reversible redox waves involving 15 electrons in the potential range of $+1.5$ to -2.0 V, where 13 reversible waves arise from 13 one-electron processes and another one involves two-electron process from two overlapping one-electron waves.

2.2 Phosphine Ligand-Substituted Triruthenium Complexes

The P donors in phosphines have strong σ-donation ability to the metal ions as well as moderate π-back bonding character. At the beginning of the 1970s, PPh_3-containing oxo-centered $Ru_3^{III,III,II}$ complexes $[Ru_3O(OAc)_3(PPh_3)_3]$ and $[Ru_3O(OAc)_6(CO)(PPh_3)_2]$ had already been isolated and structurally characterized

[1–4]. Analogous to N-heterocyclic ligands, P donors in phosphines afford favorable affinity to oxo-centered triruthenium clusters through axial coordination. These phosphine ligands have been demonstrated to afford molecular orbitals with suitable energies to overlap with those of the attached oxo-centered triruthenium cluster. By proper selection of P donor ligands, including diphosphines and polyphosphines, cluster-cluster electronic interaction between Ru_3 cluster units are tunable.

2.2.1 Diphosphine-Substituted Triruthenium Complexes

A series of oxo-centered triruthenium–acetate cluster monomers and dimers with diphosphine ligands were prepared by reactions of $Ph_2PNHPPh_2$, $Ph_2P(CH_2)_nPPh_2$ ($n = 1$–5), $trans$-$Ph_2PCH = CHPPh_2$, $Ph_2PC \equiv CPPh_2$, or $Ph_2PCp_2FePPh_2$ with one or two equivalents of methanol-coordinated $Ru_3^{III,III,III}$ precursors $[Ru_3O(OAc)_6 (L)_2(CH_3OH)]^+$ (L = dmap, py, abco) at ambient temperature [25–27]. One- or two-electron reduced tri- and hexanuclear ruthenium complexes with intracluster mixed valences were also accessed by chemical reduction. It is demonstrated that intercluster interaction is more or less operating between two oxo-centered triruthenium cluster units through various diphosphine ligands. As revealed by the amount of $\Delta E_{1/2}$ from redox wave splitting, cluster–cluster interaction in diphosphine-linked Ru_3 cluster dimers is greatest for the redox process involving $Ru_3^{III,III,III}/Ru_3^{III,III,II}$ states. This is analogous to that found in 1,4-phenylene diisocyanide-linked neutral dimers $\{Ru_3O(OAc)_6(L)_2\}_2(\mu\text{-}CNC_6H_4NC)$ (L = py, dmap) [28], but is in striking contrast to the general observation in pyrazine-linked oxo-centered Ru_3 cluster dimers, which afford progressively enhanced cluster–cluster interactions with increase of the formal oxidation states or electron contents [16, 17].

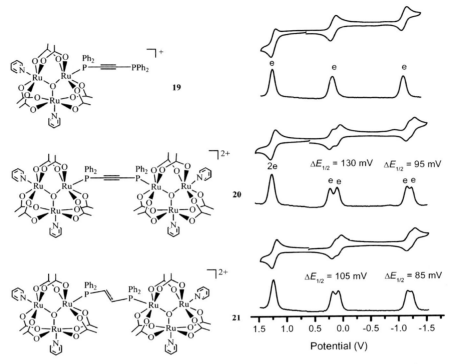

Fig. 1 Cyclic and differential voltammograms of Ph₂PC ≡ CPPh₂ (**19** and **20**) and *trans*-Ph₂PCH = CHPPh₂ (**21**) triruthenium cluster complexes recorded in 0.1 M dichloromethane solution of (Bu₄N)(PF₆). The scan rate is 100 mV s⁻¹ for CV and 20 mV s⁻¹ for DPV

The intercluster interaction between two triruthenium moieties was tuned by modification of the ancillary ligands bonded axially to the triruthenium cluster cores as well as by changing the linking spacer between two P donors in the diphosphines. It has been demonstrated that substitution of the axially coordinated pyridine (py) with *N,N*-dimethylaminopyridine (dmap) induces enhanced cluster-cluster electronic coupling. As indicated in Fig. 1, intercluster electronic coupling between two triruthenium moieties through ethynyl-spaced diphosphine (Ph₂PC ≡ CPPh₂) is obviously stronger than that mediated by ethynyl-linked ligand (*trans* − Ph₂PCH = CHPPh₂) [25].

It is intriguing that redox wave splitting ($\Delta E'_{1/2} = 0.186$ V and $\Delta E'_{1/2} = 0.123$ V) in Ph₂PNHPPh₂-linked complex [{Ru₃O(OAc)₆(py)₂}₂{μ-Ph₂PNHPPh₂}]²⁺ (**22**) is more remarkable than that in Ph₂PCH₂PPh₂-linked compound **23** ($\Delta E'_{1/2} = 0.145$ V and $\Delta E''_{1/2} = 0.118$ V) [27]. For a series of diphosphine-linked dimeric complexes [{Ru₃O(OAc)₆(py)₂}₂{μ-Ph₂P(CH₂)ₙPPh₂}]²⁺ ($n = 1$–5), they exhibit a two-electron oxidation wave composed of two closely spaced one-electron processes and two separate one-electron waves in the positive region, together with two distinctly spaced one-electron waves in the negative region. As depicted in Fig. 2,

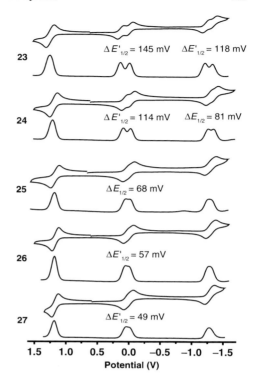

Fig. 2 Cyclic and differential voltammograms of diphosphine-linked Ru₃ cluster dimers [{Ru₃O(OAc)₆(py)₂}₂{μ-Ph₂P(CH₂)ₙPPh₂}]²⁺ (n = 1 **23**, 2 **24**, 3 **25**, 4 **26**, 5 **27**) recorded in 0.1 M dichloromethane solution of (Bu₄N)(PF₆). The scan rate is 100 mV s⁻¹ for CV and 20 mV s⁻¹ for DPV

both potential differences $\Delta E'_{1/2}$ ($\Delta E'_{1/2} = E_{1/2}^{(2+/+)} - E_{1/2}^{(+/0)}$) of two successive one-electron processes in the positive region and $\Delta E''_{1/2}$ ($\Delta E''_{1/2} = E_{1/2}^{(0/-)} - E_{1/2}^{(-/2-)}$) in the negative region reduce sharply with increase of methylene number in diphosphines Ph₂P(CH₂)ₙPPh₂ (n = 1–5) from **23** (n = 1) to **27** (n = 5). Consequently, redox wave splitting between two triruthenium cluster centers in diphosphine-linked Ru₃ cluster dimers is highly sensitive to the length and electronic nature of bridging diphosphines.

The redox wave splitting in diphosphine-linked triruthenium cluster dimers likely arises from two possible factors. One is ligand-mediated cluster–cluster interaction through an orbital pathway and determined by the extent of cluster–π*(ligand)-cluster mixing and the other is through-space electrostatic effect. Since the potential separations $\Delta E_{1/2}$ change with the electron contents or the oxidation states in the triruthenium clusters, cluster–cluster electronic interactions should make a major contribution to the redox wave splitting $\Delta E_{1/2}$.

2.2.2 Polyphosphine-Substituted Triruthenium Complexes

Reactions of (Ph₂PCH₂)₃CCH₃, PhP(CH₂CH₂PPh₂)₂ and P(CH₂CH₂PPh₂)₃ with three or four equivalents of oxo-centered triruthenium precursor complex **1** caused isolation of Ru₃ cluster dimer [{Ru₃O(OAc)₆(py)₂}₂{μ-Ph₂PCH₂)₃CCH₃}]²⁺

(28), trimer $[\{Ru_3O(OAc)_6(py)_2\}_3\{\mu_3\text{-}PPh(CH_2CH_2PPh_2)_2\}]^{3+}$ **(29)**, and tetramer $[\{Ru_3O(OAc)_6(py)_2\}_4\{\mu_4\text{-}P(CH_2CH_2PPh_2)_3\}]^{4+}$ **(30)**, respectively [27]. Interestingly, one of three P donors in 1,1,1-tri(diphenyl–phosphinomethyl)ethane is not bound to the Ru_3 unit due probably to the steric effect, inducing isolation of **28** as a dimer instead of the designed trimeric species.

The ^{31}P NMR spectra of **28–30** showed signals characteristic of their structural features. Dimeric species **28** with 1,1,1-tri(diphenylphosphinomethyl)ethane affords two singlets at 23.0 and -28.6 ppm due to the coordinated and noncoordinated P donors, respectively. For trimeric species **29**, two singlets occur at 24.3 and 14.4 ppm, ascribed to two classes of P donors in bis[2-(diphenylphosphino)ethyl]phenylphosphine. Because of two types of P donors in tri[2-(diphenylphosphino)ethyl]phosphine, tetrameric species **30** exhibits two singlets at 16.4 and 24.0 ppm.

Fig. 3 Cyclic and differential voltammograms of polyphosphine-linked Ru₃ cluster dimer **28**, trimer **29** and tetramer **30**, recorded in 0.1 M dichloromethane solution of (Bu₄N)(PF₆). The scan rate is 100 mV s⁻¹ for CV and 20 mV s⁻¹ for DPV

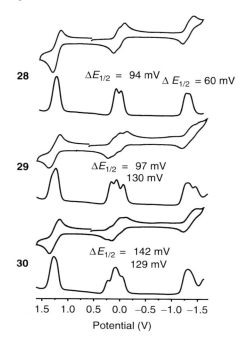

Electrochemical studies revealed that dimeric **28** exhibits redox splitting with $\Delta E_{1/2} = 0.094$ V for the $\text{Ru}_3^{\text{III,III,III}}/\text{Ru}_3^{\text{III,III,II}}$ process and 0.06 V for the $\text{Ru}_3^{\text{III,III,II}}/\text{Ru}_3^{\text{III,II,II}}$ process, obviously more remarkable than those in Ph₂P(CH₂)₃PPh₂-linked dimeric species **25** [27]. As indicated in Fig. 3, trimeric complex **29** exhibits six reversible or quasi-reversible redox waves in the range of +1.5 to −1.5 V. Three distinct one-electron redox waves occur at $E_{1/2}^{(3+/2+)} = +0.159$, $E_{1/2}^{(2+/+)} = +0.062$, and $E_{1/2}^{(+/0)} = -0.068$ V due to stepwise one-electron oxidation of three $\text{Ru}_3\text{O}(\text{OAc})_6(\text{py})_2^{\text{III,III,II}}$ into $\text{Ru}_3\text{O}(\text{OAc})_6(\text{py})_2^{\text{III,III,III}}$ species. Since cluster–cluster interaction through the shorter Ru₃–P(CH₂)₂P–Ru₃ array is stronger that through the longer Ru₃–P(CH₂)₂P(CH₂)₂P–Ru₃ pathway in trimeric species **29**; redox wave splitting for the former ($\Delta E_{1/2} = 0.130$ V) is obviously more pronounced than for the latter ($\Delta E_{1/2} = 0.097$ V). In the negative region, less noticeable wave splitting is observed due to stepwise one-electron reduction of three $\text{Ru}_3\text{O}(\text{OAc})_6(\text{py})_2^{\text{III,III,II}}$ into $\text{Ru}_3\text{O}(\text{OAc})_6(\text{py})_2^{\text{III,II,II}}$ species. The cyclic voltammogram (CV) and differential pulse voltammogram (DPV) of tetramer **30** (Fig. 3) afford five reversible or quasi-reversible redox waves. As there exist two different cluster–cluster coupling pathways with different distances in this tetrameric species, the shorter Ru₃–P(CH₂)₂P–Ru₃ array is much more efficient for mediating the cluster–cluster interaction than that of the longer Ru₃–P(CH₂)₂P(CH₂)₂P–Ru₃.

3 Triruthenium Complexes from Bridging Ligand Substitution

In contrast with numerous studies on axial substitution reactivity, related investigations on the bridging acetate substitution are in their infancy stage. The six bridging acetates in oxo-centered triruthenium cluster core $Ru_3O(OAc)_6$ are significantly stabilized by six-membered rings formed by six $Ru_2O(OAc)$ moieties. Ligand substitution of these bridging acetates is much more difficult than that of the axial ligands. Although such reactions could likely be promoted by raising reaction temperature, the oxo-centered triruthenium cluster core is always damaged during the reactions. Attempting to displace the bridging acetates in a mild condition, recent studies in our group indicated substitution of one of the six bridging acetates in $Ru_3(\mu_3\text{-}O)(\mu\text{-}OAc)_6$ core is indeed feasible by some N-heterocyclic ligands with low π^* level orbitals [8, 29–31].

3.1 Triruthenium Complexes with Ortho-Metallated Ligands

By reactions of methanol-coordinated parent triruthenium species **2** with a series of N-heterocyclic ligands, the *ortho* C–H bonds in these N-heterocycles can be activated and cleaved, inducing isolation of oxo-centered triruthenium derivatives with one of the bridging acetates substituted by an N-heterocyclic ligand [8,29,30]. Here, the parent cluster core $Ru_3O(OAc)_6$ is converted to $Ru_3O(OAc)_5(L)$ (L = *ortho*-metallated ligand) containing an *ortho*-metallated N-heterocyclic ligand.

3.1.1 Triruthenium Complexes with Ortho-Metallated 2,2′-bipyridine or 1,10-Phenanthroline

The reactions of methanol-coordinated triruthenium complex **2** with diimine ligands such as 2,2′-bipyridine and 1,10-phenanthroline at ambient temperature induced isolation of a series of oxo-centered triruthenium derivatives $[Ru_3O(OAc)_5\{\mu\text{-}\eta^1(C),\eta^2(N,N)\text{-bipyridine}\}(py)_2]^+$ (bipyridine = dbbpy **31**; dmbpy **32**; bpy **33**; Br$_2$bpy **34**; phen **35**) containing an *ortho*-metallated bipyridine [8]. Formation of triruthenium derivatives **31–35** is involved in substitution of the coordinated methanol as well as one of the six bridging acetates in the precursor complex **2** by an *ortho*-metallated bipyridine in a $\mu\text{-}\eta^1(C),\eta^2(N,N)$ bonding fashion.

The triruthenium derivatives **31–35** show characteristic intracluster charge transfer (IC) absorptions in the visible to near-infrared region (600–1000 nm) and cluster-to-ligand charge transfer (CLCT) transitions at 320–450 nm. Compared with the low energy bands in $[Ru_3^{III,III,III}]^+$ complexes **31–35**, those in the one-electron reduced neutral $[Ru_3^{III,III,II}]^0$ species are remarkably red-shifted. The decrease in energy for these transitions by one-electron reduction reflects a rise of the occupied d_π levels as the number of electrons increases. Complexes **31–35** exhibit

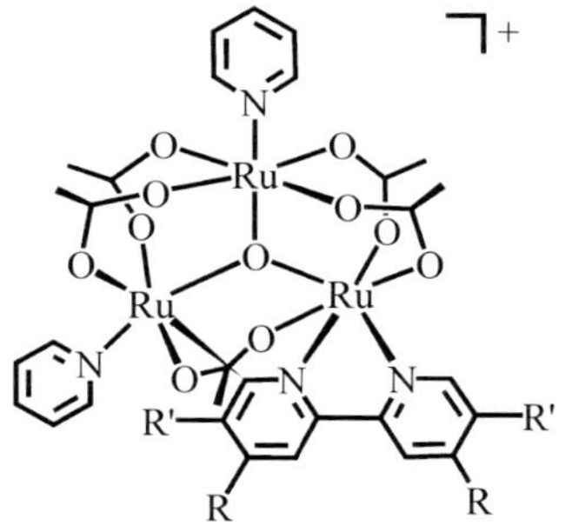

Oxo-Centered Triruthenium-Acetate Cluster Complexes

31 R = Bu^t, R' = H
32 R = CH$_3$, R' = H
33 R = H, R' = H
34 R = H, R' = Br

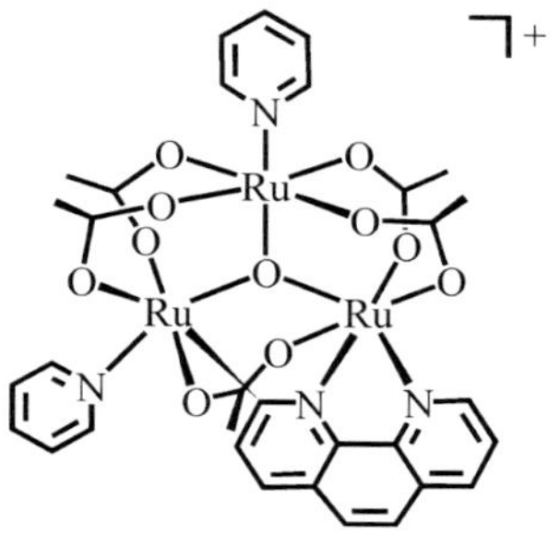

35

three reversible redox waves in the range $+1$ to -2.0 V in 0.1 M dichloromethane solution of $(Bu_4N)(PF_6)$, ascribable to successive one-electron redox processes $[Ru_3^{IV,III,III}]^{2+}/[Ru_3^{III,III,III}]^+$ $(E_{1/2}^{(2+/+)})$, $[Ru_3^{III,III,III}]^+/[Ru_3^{III,III,II}]^0$ $(E_{1/2}^{(2+/+)})$, and $[Ru_3^{III,III,II}]^0/[Ru_3^{III,II,II}]^-$ $(E_{1/2}^{(0/-)})$. Interestingly, with increase of the π-electron accepting capability in 2,2'-bipyridyl ligands, redox potentials are progressively anodic-shifted in the order **31** $\rightarrow$ **32** $\rightarrow$ **33** $\rightarrow$ **34** [8]. The stabilization against disproportionation for the oxidation states $[Ru_3^{III,III,III}]^+$ and $[Ru_3^{III,III,II}]^0$ are **31** > **32** > **33** > **34**. Consequently, it appears that introducing electron-donating substituents to the 2,2'-bipyridyl favors stabilizing the $[Ru_3^{III,III,III}]^+$ and $[Ru_3^{III,III,II}]^0$ states, whereas the electron-withdrawing substituents would destabilize it against disproportionation. Furthermore, the capability for $[Ru_3^{III,III,III}]^+$ and $[Ru_3^{III,III,II}]^0$ states against disproportionation is distinctly reduced compared with the parent triruthenium complex $[Ru_3O(OAc)_6(py)_3]^+$.

3.1.2 Tetraimine-Substituted Complexes

The catalysis of triruthenium cluster on *ortho* C–H activation and cleavage in the 2,2'-bipyridyl ligands to substitute a bridging acetate in the parent oxo-centered $Ru_3O(OAc)_6$ moiety opens up an important synthetic approach for the design of multifunctional materials composed of parent $Ru_3(\mu_3\text{-}O)(\mu\text{-}OAc)_6$ as well as derivative $Ru_3(\mu_3\text{-}O)(\mu\text{-}OAc)_5$ cluster moieties. The successful isolation of diimine-substituted oxo-centered triruthenium cluster derivatives containing 2,2'-bipyridyl ligands prompted the attempt to cleave the *ortho* C–H bonds in tetraimine or hexaimine ligands for design of dimeric or trimeric Ru_3 cluster-based species containing *ortho*-metallated bis(2,2'-bipyridyl) or tri(2,2-bipyridyl) ligands [29, 30].

2,2′-Bipyrimidine-Substituted Complexes

36 n = 1
36a n = 0

37 n = 2
37a n = 1
37b n = 0

2,2′-Bipyrimidine triruthenium derivative $[Ru_3O(OAc)_5(py)_2(\mu\text{-}\eta^1(C),\eta^2(N,N)\text{-bpym})]^+$ (**36**) could be prepared by reaction of methanol-coordinated triruthenium complex **2** with 1.5 equivalents of bpym at ambient temperature [29]. Reduction of **36** by addition of aqueous hydrazine gave one-electron reduced neutral species $[Ru_3O(OAc)_5(py)_2(\mu\text{-}\eta^1(C),\eta^2(N,N)\text{-bpym})]$ (**36a**). Reaction of bpym with 2.4 equivalents of **2**, however, induced isolation of dimeric species $[\{Ru_3O(OAc)_5(py)_2\}_2(\mu_4\text{-}\eta^1(C),\ \eta^1(C),\ \eta^2(N,N),\eta^2(N,N)\text{-bpym})]^{2+}$ (**37**) and intercluster mixed-valence complex $[\{Ru_3O(OAc)_5(py)_2\}_2(\mu_4\text{-}\eta^1(C),\eta^1(C),\ \eta^2(N,N),\eta^2(N,N)\text{-bpym})]^+$ (**37a**) during chromatographic separation on an alumina column. Addition of excess aqueous hydrazine to **37** or **37a** resulted in formation of reduced neutral product $[\{Ru_3O(OAc)_5(py)_2\}_2(\mu_4\text{-}\eta^1(C),\eta^1(C),\eta^2(N,N),\eta^2(N,N)\text{-bpym})]$ (**37b**). The dimeric species **37**, **37a**, and **37b** with different number of charges or formal oxidation states are readily interconvertible through chemical oxidation or reduction by addition of ferrocenium hexafluorophosphate or hydrazine.

Structural determination of the intercluster mixed-valence ($Ru_3^{III,III,III}$–bpym – $Ru_3^{III,III,II}$) complex **37a** (Fig. 4.) by X-ray crystallography revealed that relevant bond distances and angles around one Ru_3 cluster unit are the same as those around the other, demonstrating that intercluster mixed-valence system is electronically delocalized. The dianionic bpym behaves as a hexadentate bis(diimine)ligand in a $\mu_4\text{-}\eta^1(C),\eta^1(C),\eta^2(N,N),\eta^2(N,N)$ mode, chelating one ruthenium center through N,N′ donors and bound to another ruthenium center through deprotonated C donor by *ortho*-metallation. The intracluster Ru–Ru, Ru$-O_{oxo}$ and Ru$-O_{acetate}$ distances are in the normal ranges except for the Ru$-O_{acetate}$ (2.186(9) Å) trans-oriented to the Ru–C bond being elongated significantly due to the remarkable *trans* effect induced by *ortho*-metallation of bpym.

The CV and DPV of bpym-substituted triruthenium complex **36** (Fig. 5) exhibit three reversible redox waves at $+0.75$, -0.39, and -1.73 V, ascribed to redox processes $Ru_3^{IV,III,III}/Ru_3^{III,III,III}$, $Ru_3^{III,III,III}/Ru_3^{III,III,II}$ and $Ru_3^{III,III,II}/Ru_3^{III,II,II}$, respectively. For bpym-linked dimeric Ru_3 cluster species **37** (Fig. 5), the corresponding redox waves in individual triruthenium cluster units show notable splitting,

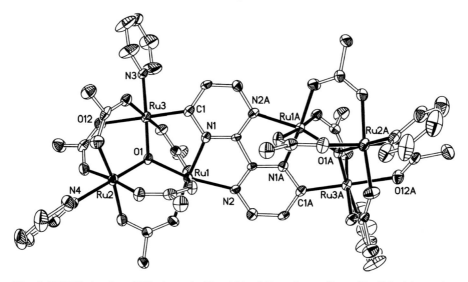

Fig. 4 ORTEP drawing (30% thermal ellipsoids) of Ru₃ cluster dimer **37a** linked by *ortho*-metallated bpym

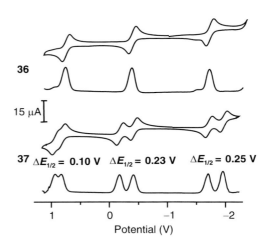

Fig. 5 Cyclic and differential voltammograms of Ru₃O cluster monomer **36** and dimer **37** with *ortho*-metallated 2,2′-bipyrimidine, recorded in 0.1 M dichloromethane solution of (Bu₄N)(PF₆). The scan rate is 100 mV s⁻¹ for CV and 20 mV s⁻¹ for DPV

which can be estimated by the potential differences $\Delta E_{1/2}$. The wave splitting $\Delta E_{1/2}$ is progressively increased with decrease of the formal oxidation states in the order $\mathrm{Ru}_3^{IV,III,III}$ (0.10 V) < $\mathrm{Ru}_3^{III,III,III}$ (0.23 V) < $\mathrm{Ru}_3^{III,III,II}$ (0.25 V). This demonstrates distinctly that more remarkable cluster–cluster interaction is operating with decrease of the oxidation states in the Ru₃O cluster moiety.

As observed in other oxo-centered triruthenium-acetate complexes, the electronic absorption spectra of **36** and dimer **37** with *ortho*-metallated 2,2′-bipyrimidine are featured by absorptions in the UV region from ligand-centered transitions, intense bands (330–410 nm) in the near UV region due to cluster-to-ligand charge transfer (CLCT) transitions from the occupied d_π orbitals of triruthenium cluster to the lowest unoccupied π^* orbitals of the ligands, and broad composite bands in the visible to near-infrared region arising from intracluster charge transfer (IC) transitions. The absorptions from both CLCT and IC transitions are sensitive to the formal oxidation states or charges in the Ru_3 cluster moiety. With decrease of the oxidation states in Ru_3 cluster dimers, both the CLCT and IC absorptions are progressively red-shifted according to **37** (350 nm) → **37a** (382 nm) → **37b** (404 nm) for CLCT absorption, and **37** (727 nm) → **37a** (752 nm) → **37b** (814 nm) for IC absorption. The possible IVCT band in the 1+ complex **37a** with intercluster mixed-valence, however, is unobserved in the scanning range 200–3000 nm, likely owing to the severe overlapping by intense IC bands.

Bis(2,2′-bipyridyl)-Substituted Complexes

Reaction of triruthenium precursor **2** with one equivalent of bis(2,2′-bipyridyl) ligand gave monomeric species **38** or **39** with *ortho*-metallated bpyC≡Cbpy or bpyC≡C–C≡Cbpy, respectively [30]. The dimeric species **40** or **41** is then

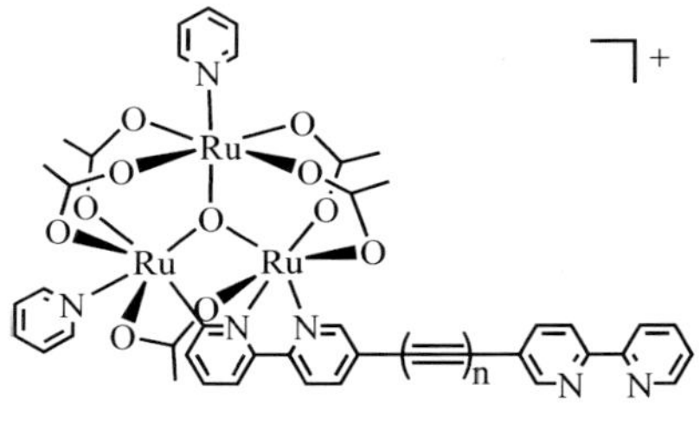

38 n = 1
39 n = 2

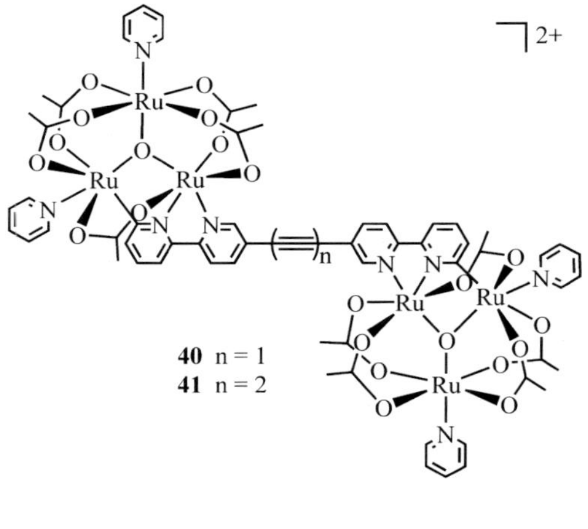

40 n = 1
41 n = 2

Oxo-Centered Triruthenium-Acetate Cluster Complexes

accessible by reaction of **38** or **39** with one equivalent of methanol-containing complex **2**, respectively. Surprisingly, the triruthenium cluster dimer **40** or **41** can not be attained by direct reaction of bpyC≡Cbpy or bpyC≡C–C≡Cbpy with two equivalents of triruthenium precursor **2**. Structural analysis of **39** by X-ray crystallography demonstrated unambiguously that the 2,2′-bipyridyl in bpyC≡C–C≡Cbpy is indeed *ortho*-metallated to the ruthenium centers.

Both monomeric complexes **38** and **39** exhibit three reversible redox waves due to successive one-electron redox processes $[Ru_3^{IV,III,III}]^{2+} / [Ru_3^{III,III,III}]^{+}$, $[Ru_3^{III,III,III}]^{+} / [Ru_3^{III,III,II}]^{0}$, and $[Ru_3^{III,III,II}]^{0} / [Ru_3^{III,II,II}]^{-}$. For dimeric species **40** or **41** containing two identical $Ru_3O(OAc)_5(py)_2$ clusters linked by bis(*ortho*-metallated) bpyC≡Cbpy or bpyC≡C–C≡Cbpy, three reversible redox waves occur nearly at the same potentials as those in monomeric species **38** or **39**, but the current strength is almost doubled for each redox wave. This suggests that cluster–cluster interaction between two identical triruthenium clusters linked by bis(*ortho*-metallated) bpyC≡Cbpy or bpyC≡C–C≡Cbpy is too weak to be resolvable by electrochemical studies. Therefore, both ethynyl (–C≡C–) and 1,3-butadiynyl (–C≡CC≡C–) linkers between two 2,2′-bipyridyls are disadvantageous for mediation of cluster-cluster electronic interactions.

2,2′-Bipyrazine-Substituted Complexes

The 2,2′-bipyrazine (bpz)-substituted Ru_3 cluster monomer **42**, dimer **43**, and trimer **44** could be accessed by reaction of triruthenium precursor **2** with different amounts of 2,2′-bipyrazine [30]. The trimeric species **44** containing two parent $Ru_3O(OAc)_6(py)_2^{III,III,III}$ and one derivate $Ru_3O(OAc)_5(py)_2^{III,III,III}$ units could be directly prepared by reaction of 3.8 equivalents of **2** with 2,2′-bipyrizine. It is also accessible by reaction of dimeric species **43** with 1.8 equivalents of **2**. The bpz adopts $\eta^1(N),\mu-\eta^1(N),\eta^1(N)$ and $\mu_4-\eta^1(N),\eta^1(N),\eta^1(C),\eta^2(N,N)$ bonding modes in **42**, **43**, and **44**, respectively. Reduction of 3+ trimer **44** by addition of aqueous hydrazine allowed isolation of neutral intracluster mixed-valence species **44b** containing three $Ru_3^{III,III,II}$ units. Oxidation of **44b** with two equivalents of ferrocenium hexafluorophosphate induced isolation of 2+ intercluster heterovalent species **44a** that contains one $Ru_3^{III,III,II}O(OAc)_5(py)_2$ and two $Ru_3^{III,III,III}O(OAc)_6(py)_2$ moieties.

Monomeric η^1(N)-bpz species **42** exhibits three reversible redox waves due to successive one-electron redox processes $[Ru_3^{IV,III,III}]^{2+}/[Ru_3^{III,III,III}]^+$, $[Ru_3^{III,III,III}]^+/[Ru_3^{III,III,II}]^0$ and $[Ru_3^{III,III,II}]^0/[Ru_3^{III,II,II}]^-$. For dimeric complex **43** containing two identical $Ru_3O(OAc)_6(py)_2$ clusters linked by a μ-η^1(N),η^1(N)-bpz, the redox wave involving the $[Ru_3^{III,III,II}]^0/[Ru_3^{III,II,II}]^-$ process for individual Ru_3 cluster unit shows appreciable splitting to afford two separate one-electron waves with $\Delta E_{1/2}$=0.08 V, although wave splitting involving $[Ru_3^{IV,III,III}]^{2+}/[Ru_3^{III,III,III}]^+$ and $[Ru_3^{III,III,III}]^+/[Ru_3^{III,III,II}]^0$ processes is not resolvable by electrochemical studies. For trimeric species **44** containing two identical $Ru_3O(OAc)_6(py)_2$ and one $Ru_3O(OAc)_5(py)_2$ moieties, eight reversible redox waves (Fig. 6) occur in the range of +1.5 to −2.5 V. Three redox waves are far from those in **42** and **43**, ascribed to successive one-electron redox processes $Ru_3^{IV,III,III}/Ru_3^{III,III,III}$, $Ru_3^{III,III,III}/Ru_3^{III,III,II}$, and $Ru_3^{III,III,II}/Ru_3^{III,II,II}$ from the $Ru_3O(OAc)_5(py)_2$ moiety with *ortho*-metallated 2,2′-bipyrazine. The other five redox waves are likely to originate from redox interactions of two identical $Ru_3O(OAc)_6(py)_2$ moieties, in which the redox waves involving $Ru_3^{III,III,III}/Ru_3^{III,III,II}$ and $Ru_3^{III,III,II}/Ru_3^{III,II,II}$ processes for individual Ru_3 units show distinct wave splitting with $\Delta E_{1/2} = 0.11$ V and $\Delta E'_{1/2} = 0.16$ V, respectively. This demonstrated again that cluster-cluster interaction is obviously

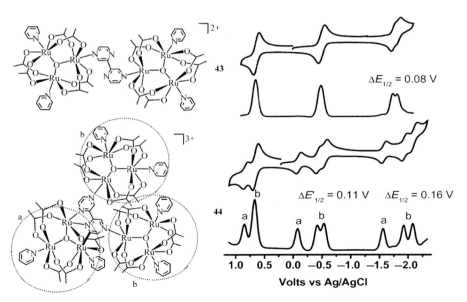

Fig. 6 Plots of cyclic and differential pulse voltammograms for bpz-linked species **43** and **44** in 0.1 M dichloromethane solution of $(Bu_4N)(PF_6)$. The scan rates are $100\,mV\,s^{-1}$ for CV and $20\,mV\,s^{-1}$ for DPV

Oxo-Centered Triruthenium-Acetate Cluster Complexes 115

enhanced with decrease of the formal oxidation states because the extent of cluster–π^*(bpz) mixing is intensified with increase of the electron content in the $Ru_3O(OAc)_6(py)_2$cluster moieties. Interestingly, redox wave splitting for two identical $Ru_3O(OAc)_6(py)_2$ is obviously more pronounced in trimeric species **44** ($\Delta E_{1/2} = 0.11$ and 0.16 V) compared with that in dimeric species **43** ($\Delta E_{1/2} < 0.05$ and $= 0.08$ V). *Ortho*-metallation of 2,2$'$-bipyrazine in trimeric complex **44** induces better coplanarity of the two pyrazine rings so as to favor cluster–π^*(ligand)–cluster orbital mixing. As a result, *ortho*-metallation of 2,2$'$-bipyrazine causes a noticeably enhanced cluster–cluster interaction.

Pyrazino[2,3-f]quinoxaline-Substituted Complexes

In contrast with formation of three types of bpz-substituted Ru_3 cluster species, reactions of **2** with pyq induced isolation of monomer **45** and trimer **46** containing *ortho*-metallated pyq depending on the reaction conditions [30]. Reduction of the 3+ trimeric complex **46** by addition of aqueous hydrazine gave neutral species **46a**. Oxidation of **46a** by addition of two equivalents of ferrocenium hexafluorophosphate afforded 2+ intercluster heterovalent complex**46b** containing two $Ru_3O(OAc)_6(py)_2^{III,III,III}$ and one $Ru_3O(OAc)_5(py)_2^{III,III,II}$ moieties.

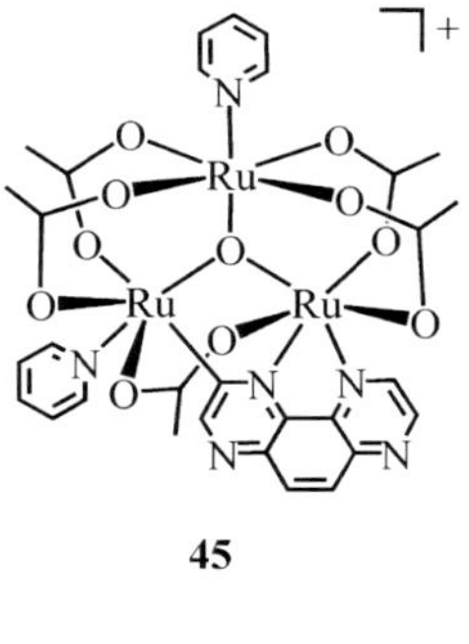

45

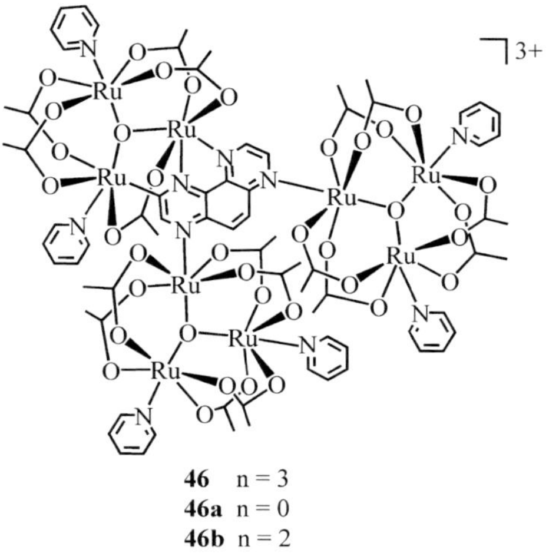

46   n $= 3$
46a n $= 0$
46b n $= 2$

As shown in Fig. 7, the electronic spectrum of 2+ intercluster heterovalent trimeric complex **46b** includes IC absorption envelopes characteristic of both $[Ru_3O]^0$ and $[Ru_3O]^+$ moieties, which are only slightly perturbed compared with those in the corresponding neutral (**46a**) and 3+ complexes (**46**). Three reversible redox waves occur at $+0.89$, -0.29, and -1.57 V in the CV and DPV of monomeric complex **45** with *ortho*-metallated pyq, ascribed to successive one-electron processes $Ru_3^{IV,III,III}/Ru_3^{III,III,III}$, $Ru_3^{III,III,III}/Ru_3^{III,III,II}$, and

Fig. 7 UV–vis-nIR absorption spectra of 3+ trimer **46** (*line*), neutral trimer **46a** (*dots*) and 2+ trimer **46b** (*dashes*) containing *ortho*-metallated pyq, measured in dichloromethane solutions

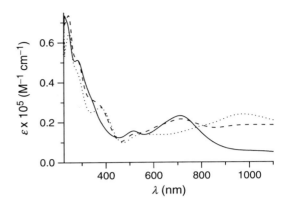

Ru$_3^{III,III,II}$/Ru$_3^{III,II,II}$, respectively. For the trimeric complex **46**, eight reversible redox waves are observed in the range of +1.5 to −2.5 V, in which the waves at +0.89 (Ru$_3^{IV,III,III}$/Ru$_3^{III,III,III}$), −0.04 (Ru$_3^{III,III,III}$/Ru$_3^{III,III,II}$), and −1.50 V (Ru$_3^{III,III,II}$/Ru$_3^{III,II,II}$) are from Ru$_3$O(OAc)$_5$(py)$_2$ unit and other five belong to two identical Ru$_3$O(OAc)$_6$(py)$_2$ subunits that exhibit distinctly redox splitting. The redox splitting involving the Ru$_3^{III,III,II}$/Ru$_3^{III,II,II}$ ($\Delta E'_{1/2}$=0.13 V) process for individual Ru$_3$ units is larger that that involving Ru$_3^{III,III,III}$/Ru$_3^{III,III,II}$ ($\Delta E_{1/2}$=0.11 V), suggesting again that cluster–cluster interaction is enhanced with decrease of the formal oxidation states in the Ru$_3$ cluster moiety.

3.2 Azo–Aromatic Ligand-Substituted Triruthenium Complexes

Analogous to dimine and tetraimine ligands, azo–aromatic ligands such as 2,2′-azobispyridine (abpy) and 2,2′-azo-bis(5-chloropyrimdine) (abcp) are π-delocalized bis-bidentate ligands with very low π* level orbitals. Reaction of oxo-centered Ru$_3^{III,III,III}$ precursor **2** with one equivalent of abpy or abcp induced isolation of Ru$_3$ cluster derivatives [Ru$_3$O(OAc)$_5${μ-η^1(N),η^2(N,N)-L}(py)$_2$]$^+$ (L = abpy **47**, abcp **48**) with formal oxidation state III,III,II [31]. Reduction of **48** by hydrazine induces isolation of one-electron reduced neutral Ru$_3^{III,II,II}$ product [Ru$_3$O(OAc)$_5${μ-η^1(N),η^2(N,N)-abcp}(py)$_2$] (**48a**). While both **47** and **48** with formal oxidation state III,III,II are diamagnetic, the room-temperature magnetic moment of Ru$_3^{III,II,II}$ species **48a** is 1.84 μ$_B$, a little higher than the spin-only value (1.73 μ$_B$) for a single unpaired electron.

From structural characterization of **48** by X-ray crystallography [31], it is suggested that formation of stable Ru$_3^{III,III,II}$ cluster derivative **47** or **48** is involved in substitution of the axially coordinated methanol as well as one of the six bridging acetates in the Ru$_3^{III,III,III}$ precursor **2** by an abpy or abcp, in which formal oxidation state of the triruthenium species is converted from III,III,III to III,III,II.

Oxo-Centered Triruthenium-Acetate Cluster Complexes

47 **48** n = 1
      48a n = 0

It appears that abpy- or abcp-substituted oxo-centered triruthenium derivatives exhibit a high stabilization on low-valence III,III,II and III,II,II species, which are usually unavailable through axial ligand substitution. The abpy or abcp exhibits a $\mu\text{-}\eta^1(N),\eta^2(N,N)$ bonding mode, chelating one ruthenium center via azo N and pyridyl/pyrimidine N donors as well as bound to another ruthenium center via the other pyridyl/pyrimidine N donor.

In striking contrast with the slight influence on spectroscopic and redox properties caused by *ortho*-metallation of 2,2'-bipyridyl ligands [8], substitution of a bridging acetate by abpy or abcp modifies dramatically the electronic features and redox properties in the oxo-centered triruthenium derivatives **47** and **48** [31]. Five distinct reversible or quasi-reversible redox waves are observed for both **47** and **48** in the potential range of $+1.5$ to $-2.0\,V$, in which the one at $E_{1/2} = -1.33\,V$ for **2** and $-0.99\,V$ for **3** are from reduction of abpy or abcp ligand, whereas the other four belong to triruthenium cluster-based $Ru_3^{III,III,IV}/Ru_3^{III,III,III}$, $Ru_3^{III,III,III}/Ru_3^{III,III,II}$, $Ru_3^{III,III,II}/Ru_3^{III,II,II}$, and $Ru_3^{III,II,II}/Ru_3^{II,II,II}$ processes. The corresponding redox potentials in **48** show 0.12–0.30 V anodic shift compared with those in **47**, probably due to the better π-accepting ability for pyrimidime-containing abcp than that for pyridyl-containing abpy.

As shown in Fig. 8, triruthenium-based redox potentials in **48** are anodic-shifted by 0.69 V for the redox process $Ru_3^{III,III,IV}/Ru_3^{III,III,III}$, 0.99 V for $Ru_3O^{III,III,III}/Ru_3O^{III,III,II}$ and 1.25 V for $Ru_3^{III,III,II}/Ru_3^{III,II,II}$ compared with the corresponding values in the parent complex $[Ru_3O(OAc)_6(py)_3]^+$. Obviously, with decrease of the formal oxidation states in the triruthenium cluster cores, the anodic shifts of redox potentials are progressively enhanced. As a result, the low-valence $Ru_3^{III,III,II}$ and $Ru_3^{III,II,II}$ species are stabilized and accessible at ambient temperature. Since the corresponding redox potentials in **47** or **48** are obviously more positive than those in $Ru_3O(OAc)_6(py)_2(CO)$ and $Ru_3O(OAc)_6(py)_2(CNR)_3$, this demonstrates clearly that abpy and abcp are even more efficient for stabilization of low-valence in the oxo-centered triruthenium complexes than those of other ligands including CO and isocyanide. Consequently, using π-delocalized neutral ligands to substitute a negatively charged acetate in the parent $Ru_3O(OAc)_6$ core modifies significantly the triruthenium-based redox potentials. In contrast, substituting one

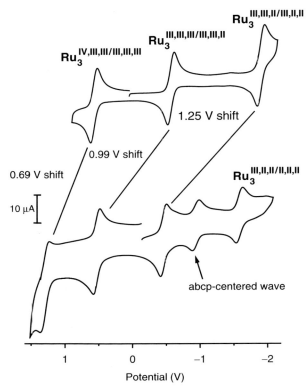

Fig. 8 Plots of cyclic voltammograms of abcp-substituted triruthenium species **48** and the parent triruthenium complex [Ru$_3$O(OAc)$_6$(py)$_3$]$^+$ in chloromethane solution of (Bu$_4$N)(PF$_6$), showing anodic shifts of redox potentials in **48** relative to those in [Ru$_3$O(OAc)$_6$(py)$_3$]$^+$

of the negatively charged acetates by a deprotonated diimine ligand through *ortho*-metallation causes little influence on the electronic and redox characteristics of the oxo-centered triruthenium species.

4 Concluding Remarks

Oxo-centered triruthenium–acetate complexes represent a class of unusual metal cluster complexes with electronically delocalized structures. These species with favorable π-delocalization exhibit a wide range of redox states by gaining or losing electrons from a series of delocalized intracluster levels. Ligand substitution affords a feasible synthetic strategy for accessing a wide range of oxo-centered triruthenium derivatives by displacing axial coordination sites or/and bridging acetates with various ligands. Alternately, ligand substitution gives an excellent means of controlling the chemical and physical properties in the oxo-centered triruthenium cluster species by elaborately designed synthetic routes. Introducing substituents on the axial or bridging ligands with different electronic effects gives a feasible approach to tuning the redox levels and electronic interaction between two or more triruthenium cluster subunits. Generally speaking, electron-donating substituents favor cluster-cluster electronic interaction whereas electron-withdrawing substituents reduce the intercluster interaction.

Although triruthenium-based monomeric and oligomeric species derived from axial ligand substitution have been investigated for more than three decades, ligand substitution of the bridging acetate has been achieved in recent years. Substitution of a bridging acetate in the parent $Ru_3O(OAc)_6$ core by an anionic *ortho*-metallated 2,2′-bipyridyl ligand exerts slight influence on the redox properties and spectroscopic features. In contrast, bridging ligand substitution by a neutral N-heterocyclic ligand with low level of π^* orbitals dramatically affects the electronic characteristics and redox levels in the triruthenium derivatives. Significant anodic shifts of the triruthenium-based redox potentials induced by substitution of a bridging acetate in the parent $Ru_3O(OAc)_6$ core with a neutral azo–aromatic ligand result in facile formation of low-valence $Ru_3^{III,III,II}$ and $Ru_3^{III,II,II}$ species that are usually inaccessible. The finding opens a significant approach to tuning the redox potentials of oxo-centered triruthenium cluster complexes through substitution of a bridging acetate by a neutral N-heterocyclic ligand with low level of π^* orbitals.

Current challenges in the design of oxo-centered triruthenium derivatives through ligand substitution reaction include (i) selection of wider range of π-delocalized N-heterocyclic ligands to substitute the bridging acetates, (ii) isolation of stable triruthenium species with wider ranges of formal oxidation states through bridging ligand substitution, and (iii) preparation of dimers or trimers derived from substitution of the bridging acetates by neutral bis(terdentate) or tris(terdentate) N-heterocyclic ligands to achieve better cluster-cluster interaction.

Acknowledgements We are grateful for financial support from the NSFC (Grants 20521101, 20625101 and 20773128), the 973 project (Grant 2007CB815304), the NSF of Fujian Province (Grant 2008I0027), and the fund from the Chinese Academy of Sciences (Grant KJCX2-YW-H01).

References

1. Cotton FA, Norman JG, Spencer A, Wilkinson G (1971) Chem Commun, p 967
2. Spencer A, Wilkinson G (1972) J Chem Soc Dalton Trans, p 1570
3. Cotton FA, Norman Jr JG (1972) Inorg Chim Acta 6:411
4. Spencer A, Wilkinson G (1974) J Chem Soc Dalton Trans, p 786
5. Sasaki Y (1995) J Mol Liq 65/66:253
6. Sasaki Y, Umakoshi K, Imamura T, Kikuchi A, Kishimoto A (1997) Pure Appl Chem 69:205
7. Toma HE, Araki K, Alexiou ADP, Nikolaou S, Dovidauskas S (2001) Coord Chem Rev 219–221:187
8. Chen JL, Zhang XD, Zhang LY, Shi LX, Chen ZN (2005) Inorg Chem 44:1037
9. Baumann JA, Salmon DJ, Wilson ST, Meyer TJ, Hatfield WE (1978) Inorg Chem 17:3342

10. Kido H, Nagino H, Ito T (1996) Chem Lett, p 745
11. Toma HE, Cunha CJ, Cipriano C (1988) Inorg Chim Acta 154:63
12. Abe M, Sasaki Y, Yamada Y, Tsukahara K, Yano S, Yamaguchi T, Tominaga M, Taniguchi I, Ito T (1996) Inorg Chem 35:6724
13. Abe M, Sasaki Y, Yamada Y, Tsukahara K, Yano S, Ito T (1995) Inorg Chem 34:4490
14. Ito T, Hamaguchi T, Nagino H, Yamaguchi T, Kido H, Zavarine IS, Richmond T, Washington J, Kubiak CP (1999) J Am Chem Soc 121:4625
15. Abe M, Michi T, Sato A, Kondo T, Zhou W, Ye S, Uosaki K, Sasaki Y. (2003) Angew Chem Int Ed 42:2912
16. Wilson ST, Bondurant RF, Meyer TJ, Salmon DJ (1975) J Am Chem Soc 97:2285
17. Baumann JA, Salmon DJ, Wilson ST, Meyer TJ (1979) Inorg Chem 18:2472
18. Ito T, Hamaguchi T, Nagino H, Yamaguchi T, Washington J, Kubiak CP (1997) Science 277:660
19. Yamaguchi T, Imai N, Ito T, Kubiak CP (2000) Bull Chem Soc Jpn 73:1205
20. Ito T, Imai N, Yamaguchi T, Hamaguchi T, Londergan CH, Kubiak CP (2004) Angew Chem Int Ed 43:1376
21. Londergan CH, Kubiak CP (2003) Chem Eur J 9:5962
22. Baumann JA, Wilson ST, Salmon DJ, Hood PL, Meyer TJ (1979) J Am Chem Soc 101:2916
23. Hamaguchi T, Nagino H, Hoki K, Kido H, Yamaguchi T, Breedlove BK, Ito T (2005) Bull Chem Soc Jpn 78:591
24. Toma HE, Alexiou ADP (1995) J Braz Chem Soc 6:267
25. Chen JL, Zhang LY, Chen ZN, Gao LB, Abe M, Sasaki Y (2004) Inorg Chem 43:1481
26. Chen JL, Zhang LY, Shi LX, Ye HY, Chen ZN (2005) Inorg Chim Acta 358:859
27. Chen JL, Zhang LY, Shi LX, Ye HY, Chen ZN (2006) Inorg Chim Acta 359:1531
28. Ota K, Sasaki H, Matsui T, Hamaguchi T, Yamaguchi T, Ito T, Kido H, Kubiak CP (1999) Inorg Chem 38:4070
29. Ye HY, Zhang LY, Chen JL, Chen ZN (2006) Chem Commun, p 1971
30. Dai FR, Chen JL, Ye HY, Zhang LY, Chen ZN (2008) Dalton Trans, p 1492
31. Ye HY, Dai FR, Zhang LY, Chen ZN (2007) Inorg Chem 46:6129

Struct Bond (2009) 133: 121–160
DOI:10.1007/430_2008_13
© Springer-Verlag Berlin Heidelberg 2009
Published online: 5 March 2009

New Progress in Monomeric Phthalocyanine Chemistry: Synthesis, Crystal Structures and Properties

Zhonghai Ni, Renjie Li, and Jianzhuang Jiang

Abstract In this chapter, recent progress in the synthesis, crystal structures and physical properties of monomeric phthalocyanines (Pcs) is summarized and analysed. The strategies for synthesis and modification of Pcs include axial coordination of central metal ions, peripheral substitution of Pc rings and the ionization of Pcs. The crystal structures of various typical Pcs, especially the effects of different synthetic and modification strategies on the supramolecular assemblies of Pcs via $\pi-\pi$ interactions between Pc rings, are discussed in detail. Finally, the UV–vis spectroscopic, conducting, magnetic and catalytic properties of some Pcs with crystal structures are presented briefly, and the correlations between various properties and the molecular structure discussed.

Keywords: Crystal structure · Modification · Phthalocyanine · Property · Synthesis

Contents

1 Introduction 123
2 Synthesis and Modifications of Phthalocyanines 123
3 Crystal Structures and Supramolecular Assemblies of Phthalocyanines 125
    3.1 Neutral Parent Phthalocyanines 126
    3.2 Neutral Parent In-Plane Phthalocyanines with Axial Ligands 128
    3.3 Neutral Parent Out-of-Plane Phthalocyanines 134
    3.4 Peripherally Symmetrical Substituted Phthalocyanines 138
    3.5 Peripherally Unsymmetrical Substituted Phthalocyanines 146
    3.6 Ionic Phthalocyanines 147
4 Properties of Phthalocyanines... 151
    4.1 UV–vis Spectra 151
    4.2 Conductivity 153

Z. Ni, R. Li, and J. Jiang (✉)
Department of Chemistry, Shandong University, Jinan 250100, China
e-mail: jzjiang@sdu.edu.cn

4.3	Magnetic Properties	156
4.4	Catalytic Properties	157
5	Conclusion	158
References		159

Abbreviations

ac	Acetone
AO	Amyloxy
BA	Benzylamine
BzT	Benzylthio
4-*t*BuB	4-(*Tertiary*-butyl)benzoate
BuO	Butyloxy
15-C-5	15-Crown-5
ClNP	Chloronaphthalene
4-CP	4-Cyanopyridine
DHCP	1,1-Dihexylcyclopentane
DMF	*N*,*N*-Dimethylformamide
DMPO	2,4-Dimethyl-3-pentyloxy
DMPP	3-(3,4-Dimethoxyphenyl)propanoate
DPP	2,3-Di(2-pyridyl)pyrazino
D*i*PPO	2,6-Diisopropylphenoxy
DTB	3,5-Di-[1,4,7,10-tetraoxaundecyl]benzyloxy
EE	Ethoxyethanol
EO	Ethyloxy
4-FP	4-Fluorophenyl
HT	Hexylthio
MEA	2-Methoxyethylamine
4-MP	4-Methylpyridine
MPc	Metal phthalocyanine
PF*i*P	Perfluoroisopropyl
Pc	Phthalocyanine, phthalocyaninato
PO	3-Pentyloxy
Py	Pyridine
PXX	*Peri*-xanthenoxanthene
3-TA	3-Thienylacetate
TMPO	2,2,4-Trimethyl-3-pentyloxy
TPIP	1-Triphenylphosphoranylidene-2-propanone
TPP	Tetraphenylphosphonium, tetraphenylporphyrin
TDOFe	Tridioximates macrobicyclic iron^{II} complex
TED	Triethylenediamine

1 Introduction

Since their discovery one century ago, phthalocyanines (Pcs) have attracted extensive interest in several research fields including organic chemistry, inorganic chemistry, material chemistry, biochemistry, magnetochemistry, supramolecular chemistry and molecular nanotechnology [1–5]. Owing to their excellent and useful chemical and physical properties such as optical, electrical, magnetic and catalytic properties, and their high physical and chemical stability, Pcs have become one of the most important materials both in conventional and high-tech fields [6–9]. Nowadays, Pcs are widely used as dyestuffs and pigments, photoconducting agents, photosensitizers, chemical sensors, electrochromic materials, photodynamic reagents, media for computer read/write discs, electrodes, non-linear optical materials, liquid crystals, catalysts, magnetic materials, conductors and so on [10–14]. For these purposes, more than 50,000 tons of Pcs are produced per year.

Owing to the strong intermolecular $\pi-\pi$ stacking in planar Pcs and the resulting inherent lack of solubility, in particular for the parent unsubstituted Pcs, together with the limited species of unsubstituted Pcs, the development of Pc chemistry faces austere challenges. In addition, these problems hinder deep investigation of the crystal structures of Pcs, which not only makes it difficult to establish the relationship between various physical behaviours and the solid state structures of Pcs, but also has limited, to some extent, their widespread application.

In a continuous effort to circumvent the problem of poor solubility and tune the steric effects and electronic features of Pcs, several effective strategies have been developed. As a result, in recent years many neutral Pcs containing substituents at peripheral α or β positions, or in the axial direction, as well as ionic and sandwich-type Pcs have been synthesized and their single-crystal structures resolved by X-ray diffraction analysis [15–24]. It therefore appears necessary to give a relatively comprehensive overview of the new progress in Pc chemistry. In this chapter, we summarize recent research results on the synthesis, crystal structures, and various physical properties of monomeric Pc compounds.

2 Synthesis and Modifications of Phthalocyanines

Metal-free and metal Pcs without peripherally substituted ligands and/or large axial coordinated ligands have been synthesized, often by cyclotetramerization reaction of phthalyl derivatives such as phthalonitrile, phthalic anhydride, 1,3-diiminosioindoline, phthlimide and phthalic acid [4]. As mentioned above, these common unsubstituted Pcs are highly insoluble in common organic solvents, which prevents them from being applied broadly in materials chemistry. In this section, we describe briefly several effective synthetic and modification strategies to improve the solubility and tune the steric and electronic features of Pcs (Fig. 1).

The first effective strategy is the introduction of axial assistant coordination ligands to the central metal atoms of Pcs [25–48]. The synthetic paths of this type of

Fig. 1 General strategies for the modification of Pcs

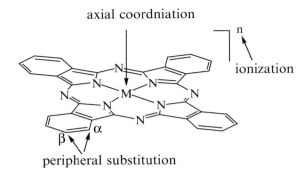

Pc are mostly via coordination reactions or axial ligands exchange reactions based on the metal Pcs themselves. This is the most successful method for modification of the parent Pcs themselves to date. Principally, the central metal ion in this type of Pc either locates with the mean $(N_{iso})_4$ plane formed by four isoindole nitrogen atoms, which is called *in-plane coordination pattern*, or significantly shifts from the mean $(N_{iso})_4$ plane, which is called *out-of-plane coordination pattern*. In principle, the additional axial ligand can be monodentate, bidentate or multidentate. The located position of axial ligands can be *trans-* or *cis*-coordinated to the central metal atom. The axial ligands employed recently include long aliphatic carboxylates, aliphatic amines and alcohol groups, which usually have a strong influence on the intermolecular $\pi-\pi$ stacking in the crystal structure and greatly enhance the solubility of Pcs [15, 16, 33–37, 46, 48]. Some small groups such as halide, azide and hydroxy anions, and neutral water, pyridine, 4-cyanopyridine, 4-methylpyridine, methanol, ethanol and even anisole molecules sometimes can also effectively weaken the strong intermolecular Pc $\pi-\pi$ interactions [25–32, 38–45]. Interestingly, the axial coordinated ligands can also link metal complexes such as porphyrin compounds and monomeric iron[II] complex, which may offer new potential developing space for functional Pc chemistry [38, 47]. The number of axial ligands and the symmetry of Pcs are mainly controlled by the nature of central metal atoms.

The second effective method is the employment of peripherally substituted ligands [49–68]. Almost all Pcs of this type are synthesized by using substituted precursors such as substituted phthalonitriles as starting materials, although in principle this type of Pc can be prepared directly by chemical modification of the Pcs themselves. Generally, the preparation conditions of peripherally functionalized Pcs based on substituted precursors are softer than those employed in the synthesis of conventional unsubstituted Pcs, because the increased solubility of starting substituted precursors and the resulting peripherally substituted Pcs allow the employment of a wide variety of solvents. The peripherally substituted ligands mainly include crown ether, alkyl, alkoxy and alkylthio groups. From the point of view of symmetry, this type of Pc includes symmetrically and unsymmetrically substituted Pcs. According to the number of substituted ligands, the symmetrically substituted Pcs can be classified as tetra-, octa- and hexadecasubstituted Pcs. Certainly, the substituted positions may be different for symmetrically tetra- and octasubstituted Pcs.

New Progress in Monomeric Phthalocyanine Chemistry

The synthetic methods of symmetrically substituted Pcs are similar to those of the parent unsubstituted Pcs. However, unsymmetrically substituted Pcs are usually prepared by statistical condensation methods, subphthalocyanine expulsion methods or cross-condensation methods.

The third effective approach is the preparation of ionic Pcs with large balanced ions [69–77]. It is well known that ionic compounds are usually more soluble than neutral compounds in most organic solvents, especially in polar and mixing solvents. Ionic Pcs include cationic and anionic Pcs. The preparation of ionic Pcs is usually by means of electrochemistry. Sometimes they can be synthesized through oxidation reactions, ion exchange reactions or ion coordination reactions.

Generally, the above three methods are effective synthetic and modification strategies for preparing soluble Pcs and tuning their properties. In the actual process of synthesizing soluble Pcs, the most effective strategy is certainly the combination of all the three methods and many Pcs have actually been synthesized based on this consideration [73–77].

3 Crystal Structures and Supramolecular Assemblies of Phthalocyanines

With the rapid development of modern analysis techniques, especially the popularization of single-crystal X-ray diffraction equipment, the origin of various physical and chemical properties, and the clear elucidation of correlations between the structure and properties of Pcs have become possibilities. Together with the maturity and diversification of their modifications, many Pcs have been synthesized and their molecular structures characterized by X-ray diffraction analysis in the past 5 years [15–77].

The basic structure of a Pc molecule is a large planar macrocycle consisting of four isoindole units circularly linked by azamethine bridges, presenting an $18-\pi$ electron aromatic cloud delocalized over the cyclic core. One common structural feature is the cavity within the centre of the molecule, which can whole or partly accommodate most elements in the periodic table. The distortion degree and molecular shape of Pc skeletons depend mainly on the size of the atoms located within the cavities, the coordination geometries of the central metal ions, the number and steric effect of axial additional ligands bound to the central metal atoms, and the peripherally substituted ligands appended to the Pc macrocycles. In order to clarify and compare reasonably the large number of Pcs with crystal structures published recently, the Pcs in this chapter will be classified on the basis of the above-discussed modification strategies, and sub-classified according to the molecular shapes (in-plane and out-of-plane) of the Pc skeletons and the molecular coordination geometries of the central metal ions [5].

Another common structural feature of Pcs is the abundant arrangements of molecules through various Pc π–π interactions. These different stacking behaviours can generally be figured out by the stacking angles of the molecules towards the stacking axes, the separation of adjacent metal ions or the parallel shift distance of two neighbouring Pc molecules. The degree of the Pc π–π interactions depends mainly on the interplanar distance (3.0–3.8 Å) and the overlapping area, which can be evaluated from calculation of the intermolecular HOMO–HOMO overlap integrals [12]. Generally, the π–π overlaps for different monomeric Pcs published recently can be divided into 14 representative types based on the largely overlapping areas and the parallel shift directions of adjacent Pc macrocycles (Fig. 2). The effect of different modifications on the crystal structures of Pcs, especially on the arrangements of Pc molecules in the solid state, will be discussed in detail.

3.1 Neutral Parent Phthalocyanines

This type of phthalocyanine includes metal-free Pcs (H_2Pc) and tetracoordinated metal Pcs (MPc) with square planar coordination surroundings (Fig. 3), which have been widely investigated and well reviewed in detail before [5]. The central metal ions include Mn^{II}, Fe^{II}, Co^{II}, Ni^{II}, Cu^{II}, Zn^{II} or Pt^{II} ions of appropriate size. Most of these Pcs crystallize in two (α and β) modifications, the β-type structure being more stable and better characterized. These Pcs usually form slipped-stacked 1D columns with the type A Pc π–π overlap (see Fig. 2 for schemes of overlap types A–N). Interestingly, the α and β crystalline polymorphs have very different molecular arrangements in the solid state, in particular in the inclination of the planar Pc molecules relative to the stacking axis, although the geometries, intramolecular bond distances and angles of the molecules in the two crystalline polymorphs are very similar. In addition, various β-type metal-free and metal Pcs have very similar structures. The adopted stable form of this type of $M^{II}Pc$ depends mainly on the intermolecular interactions in the crystals, which has been discussed in detail by Janczak and co-workers [78].

Another interesting and important problem that has been preliminarily solved recently is the structure of complex $Be^{II}Pc$ (**1**) containing the exceptionally small Be^{II} ion of group IIA. In 2006, Kubiak and co-workers [39] examined the crystal structure of the complex. The results indicate that the tetracoordinated Be^{II} atom locates at the centre of symmetry of the planar Pc(2-) moiety. The interatomic distances and angles in the complex do not show any marked deviations from the geometry characteristic of the d-forms of the tetracoordinated parent $M^{II}Pcs$. The $Be^{II}Pc$ molecules are arranged in stacks with a distance between the Pc(2-) planes of 3.4 Å, similar to other crystal structures of the β-$M^{II}Pc$ series. However, a significant difference in the intermolecular interactions between complex **1** and other d-block $M^{II}Pcs$ is found, i.e. no intermolecular weak coordination interaction between $Be^{II}Pc$ molecules in stacks is observed in **1**. The reason might be attributed to the exceptionally small size of the Be^{II} ion in complex **1**. The β-form tetracoordinated parent $M^{II}Pc$ stacking structures (type A π–π overlap) and the ionic radii of the central metals are shown in Fig. 4.

New Progress in Monomeric Phthalocyanine Chemistry

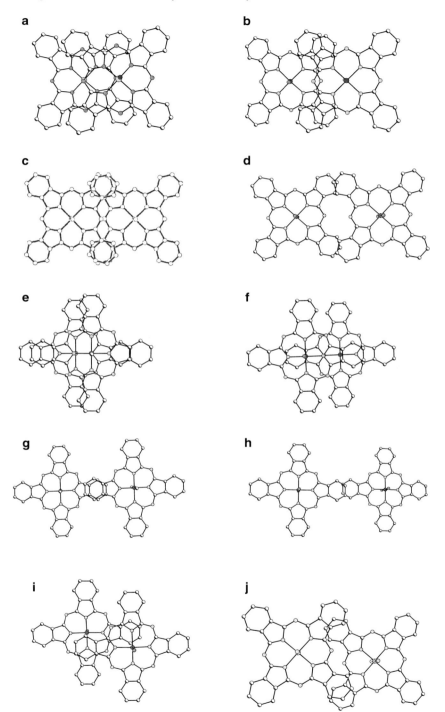

Fig. 2 Fourteen representative types (*A–N*) of Pc π–π overlaps

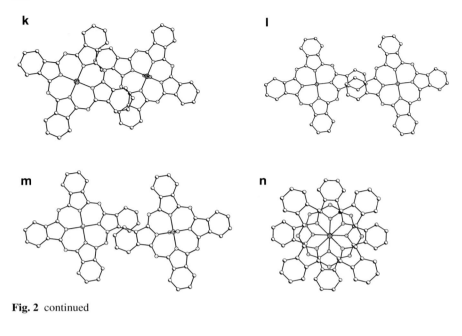

Fig. 2 continued

Fig. 3 Schematic structures of neutral parent Pcs

3.2 Neutral Parent In-Plane Phthalocyanines with Axial Ligands

Unlike the above-discussed neutral parent Pcs, the neutral parent in-plane Pcs with axial ligands have one or two additional axial ligands (Fig. 5), yielding square pyramidal and octahedral coordination geometries with large Pc(2-) or one-electron oxidized Pc(1-) radical ligands situated at the equatorial planes of the metal ions. The Pc(2-) and Pc(1-) ligands in these complexes are not usually strictly planar. However, the four isoindole coordination nitrogen atoms (N_{iso}) are nearly strictly coplanar, and the central metal ions nearly locate within the centre of the mean plane with a displacement of less than 0.1 Å from the $(N_{iso})_4$ plane. The oxidation states of the central metal ions in these Pcs can be 2, 3 or 4. For this type of MPc with two axial ligands, the two ligands mostly are the same, except in a few species. Most of these Pcs can form slipped-stacked 1D chain-like, 1D ladder-like, or even 2D sheet-like supramolecular structures. However, owing to the presence of axial additional ligands, which decrease the degree of Pc π–π overlaps between neighbouring molecules, the stacking in these crystals is no longer as compact as

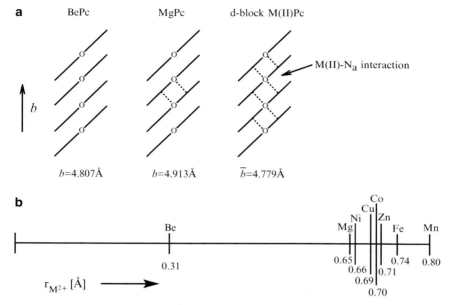

Fig. 4 Schematic drawing showing the β-form M^{II}Pc stacking structures **a** and the ionic radii of the metals **b**. Figure obtained from [39]

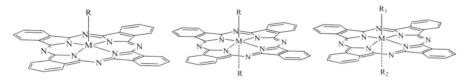

Fig. 5 Schematic structures of neutral parent in-plane Pcs with axial ligands

observed in the neutral parent Pcs. If there are strong hydrogen bonds in the crystals, the stacking arrangements become more complicated. Basically, the larger the additional axial ligand, the greater its influence on the molecular arrangements of Pcs in the crystals, until the π–π overlap between the Pc planes is prevented completely. The crystal structures and supramolecular structures of neutral parent in-plane Pcs published in the past 5 years are compiled in Table 1.

3.2.1 Pentacoordinated Phthalocyanines

Pentacoordinated Pc complexes with planar macrocycle are very scarce. Very recently, Galezowski and Kubicki [25] reported a σ-bonded Co^{III}Pc(CH$_3$CH$_2$) (**2**) with an alkyl group as axial ligand. Investigation of its crystal structure revealed that the solid consists of centrosymmetric face-to-face dimers. The cobalt atom in **2** is pentacoordinated and has a distorted square pyramidal geometry. The cobalt atom

Table 1 Neutral parent in-plane phthalocyanines

Compound	Metal dev.[a] Å	M$\cdots$M[b] Å	π–π Overlap[c]	Structure[d]	Ref.
Be^{II}Pc	0	4.807	A	1D	[39]
Co^{III}Pc(CH$_2$CH$_3$)	0.086	3.582	A, C	1D	[25]
Cr^{III}Pc(I)$_2\cdot$I$_2$	0	8.402	C	1D	[26]
Mg^{II}Pc(Py)$_2$	0	9.141	J, M	2D	[27]
Mn^{II}Pc(Py)$_2$	0	9.171	J, M	2D	[28]
Co^{II}Pc(Py)$_2$	0	9.142	J, M	2D	[29]
Fe^{II}Pc(Py)$_2$	0	9.179	J, M	2D	[29]
Fe^{II}Pc(4-CP)$_2\cdot$2(4-CP)	0	11.446	–	–	[30]
Ru^{II}Pc(4-MP)$\cdot$2CHCl$_3$	0	10.241	K	1D	[31]
Al^{III}Pc(anisole)$_2$	0	10.336	D, H	2D	[32]
Fe^{II}Pc(BA)$_2\cdot$3THF	0	10.750	–	–	[33]
Si^{IV}Pc(DTB)$_2$	0	9.208	L	1D	[34]
Si^{IV}Pc(4-tBuB)$_2$	0	8.841	L	1D	[35]
α-Si^{IV}Pc(3-TA)$_2$	0	8.243	L	1D	[35]
Si^{IV}Pc(DMPP)$_2$	0	9.730	K	2D	[35]
Si^{IV}Pc[OOC(CH$_2$)$_7$CH$_3$]$_2$	0	9.044	K, M	2D	[15]
Si^{IV}Pc[OOC(CH$_2$)$_{10}$CH$_3$]$_2$	0	9.004	J	1D	[15]
Si^{IV}Pc[OOC(CH$_2$)$_{12}$CH$_3$]$_2$	0	9.102	J	1D	[15]
Si^{IV}Pc[OOC(CH$_2$)$_{13}$CH$_3$]$_2$	0	11.030	L	1D	[15]
Si^{IV}Pc[OOC(CH$_2$)$_{20}$CH$_3$]$_2$	0	9.175	J	1D	[15]
Si^{IV}Pc(naproxene)$_2$	0.005	9.686	K	1D	[36]
β-Si^{IV}Pc(3-TA)$_2$	0.001	7.986	C	Dimer	[37]
Co^{III}Pc(CH$_3$)(Py)	0.005	8.917	J, M	1D	[25]
[Cr^{III}Pc(N$_3$)(OH)Mn^{III}TPP]$\cdot$2ClNP	0.054	7.789	C	Dimer	[38]

[a]Displacement of the central metal ions from the mean (N$_{iso}$)$_4$ plane
[b]Shortest intermolecular metal$\cdots$metal separation
[c]Type of Pc π–π overlaps as shown in Fig. 5
[d]Supramolecular structure by Pc π–π overlap

nearly locates within the centre of the mean (N$_{iso}$)$_4$ plane with a deviation of 0.086 Å towards ethyl group. In addition, the cobalt atom of one Pc faces an isoindole nitrogen atom from the neighbouring Pc macrocycle, at a separation of 3.303 Å, which is similar to those of β-form neutral tetracoordinated parent M^{II}Pcs. The intradimeric Cr$\cdots$Cr separation is 3.879 Å. It is noteworthy that although the presence of a mono-axial ethyl group does not change the π–π overlap of intradimers (type E overlap) in **2**, it significantly hinders the π–π stacking between these dimers with only relative weak type C Pc π–π overlap. The Cr$\cdots$Cr separation from two adjacent interdimers is 7.997 Å. The 1D supramolecular structure of **2** based on the Pc π–π overlaps is shown in Fig. 6.

3.2.2 Symmetrical Hexacoordinated Phthalocyanines

Symmetrical hexacoordinated Pcs with planar rings have been the focus of by several groups in the past 5 years [26–35]. In order to display clearly the influence of

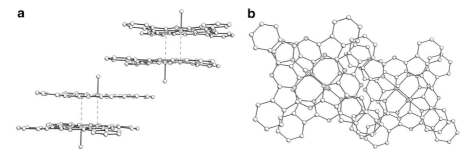

Fig. 6 Parallel **a** and perpendicular **b** views of the 1D Pc columnar arrangement in **2**; all hydrogen atoms and axial uncoordinated atoms are omitted for clarity

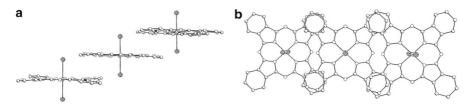

Fig. 7 Parallel **a** and perpendicular **b** views of the 1D Pc columnar arrangement in **3**; all hydrogen atoms and I$_2$ molecules are omitted for clarity

axial ligands on the basic arrangement of Pc molecules, several typical results for this type of Pc will be discussed in this section according to the size of axial ligand.

In 2002, Janczak and co-workers [26] characterized the crystal structure of Cr^{III}Pc(I)$_2$ · I$_2$ (**3**). The central Cr^{III} ion is coordinated by four N$_{iso}$ atoms from the one-electron oxidized Pc(1-) macrocyclic radical ligand and two *trans* iodine atoms, yielding a distorted CrN$_4$I$_2$ octahedral geometry. Cr^{III}(Pc)(I)$_2$ units are connected together by type C Pc π–π interactions, forming slipped-stacked 1D columnar structures as shown in Fig. 7. The Cr···Cr separation in the column is 8.402 Å. The angle between the Cr^{III}Pc plane and the stacking axis is 23.6°, which is markedly smaller than those of β-form parent MPcs without axial ligands (ca. 48.4°). These data indicate that the two axial iodine atoms effectively prevent the overlapping of Pc planes. The above columns are further linked by neutral I$_2$ molecules through the axially coordinated iodine atoms into a 2D sheet-like supramolecular structure.

From 2002 to 2007, Kubiak and Janczak [27–30] and Sun's [31] groups investigated various symmetrical MPcs with pyridine and its derivatives 4-CP and 4-MP as two axial ligands. Six complexes in this series with crystal structures are reported. The crystal structures of M^{II}Pc(Py)$_2$ (M = Mg (**4**), Mn (**5**), Co (**6**) or Fe (**7**)) complexes with axial pyridine ligands are isostructural. Another two compounds [Fe^{II}Pc(4-CP)$_2$] · 2(4-CP) (**8**) and [Ru^{II}Pc(4-MP)$_2$] · 2CHCl$_3$ (**9**) have similar molecular structures to **4**–**7**. The central metal ions in these complexes lie at the inversion centres; thus, the molecules are centrosymmetric. In the six complexes, the central metal ion and the four N$_{iso}$ atoms of the Pc(2-) ligands lie on a strict plane. The

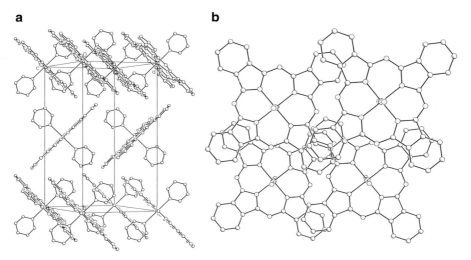

Fig. 8 Molecular arrangement in the crystal packing **a** and the single sheet structure **b**; all Py carbon atoms are omitted for clarity in **4–9**

arrangement of the $M^{II}Pc(Py)_2$ molecules in the unit cell is illustrated in Fig. 8. The four crystals are built up from isolated $M^{II}Pc(Py)_2$ molecules, which form alternating sheets in which molecules are related by a screw axis and glide plane. In one sheet the planes of Pc(2-) macrocycle are parallel to each other, while between the sheets the Pc(2-) planes are perpendicular. Within each sheet the neighbouring molecules are partly overlapped through types J and M Pc π–π stacking patterns. The shortest M···M separations in complexes **4–7** locate within a narrow range of 9.142–9.179 Å. It is noteworthy that there are no obvious Pc π–π interactions in complex **8**, and the shortest intermolecular Fe···Fe distance is relative long at 11.446 Å. Complex **9** forms 1D supramolecular polymer only via weak type K Pc π–π interactions with the shortest intermolecular Ru···Ru separation, 10.241 Å.

In 2006, Vaid's group [32] reported the structure of $Al^{III}Pc(anisole)_2$ (**10**) with Pc(1-) macrocyclic radical ligand and two axial anisole ligands. The Al^{III} ion in this complex is coordinated by four N_{iso} atoms from the equatorial Pc(1-) ligand and two oxygen atoms from two *trans* anisole ligands, giving a AlN_4O_2 octahedral geometry. The molecules of **10** stack by the combination of type D and H Pc π–π interactions, leading to a 2D sheet-like supramolecular structure (see Fig. 9). In the same year, Pregosin and Albinati [33] synthesized complex $Fe^{II}Pc(BA)_2 \cdot 3THF$ (**11**). The results indicate that the molecular structure of complex **11** consists of the $Fe^{II}(Pc)$ moiety with the two *trans*-coordinated BA ligands and three THF solvate molecules. It is noteworthy that no Pc π–π interactions are observed in this complex. In 2003, the McKeown group [34] reported the crystal structure of $Si^{IV}Pc(DTB)_2$ (**12**) with large aryl ether dendrimers as axial ligands. Complex **12** molecules slipped-stacked to form 1D supramolecular structures via type L Pc π–π overlap.

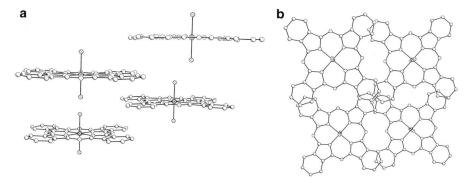

Fig. 9 Parallel **a** and perpendicular **b** views of the 2D Pc sheet arrangement in **10**; all uncoordinated atoms in axial ligands are omitted for clarity

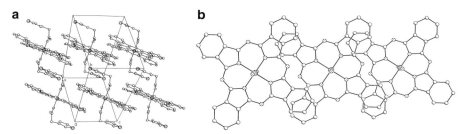

Fig. 10 Molecular arrangement in the crystal packing **a** and 1D column structure **b**; all uncoordinated atoms in axial ligands are omitted in **14**

Organic alkyl and aromatic acids are usually exploited as axial ligands because of their excellent coordination abilities for some metal atoms. In 2002, Bryce and Beeby's group [35] investigated axially substituted Si^{IV}Pcs using various rigid and flexible carboxylates as the ligands. They determined the crystal structures of three complexes Si^{IV}Pc(L)$_2$, where L indicates 4-tBuB (**13**), 3-TA (**14**) or DMPP (**15**). The molecular structures of the three complexes are similar. Owing to the steric effects of the different axial ligands, the three complexes display different molecular arrangements in the crystal packing. Complex **13** contains infinite stair-like stacks of Pc moieties by type L Pc π–π overlap. A similar Pc-stacking motif is observed in **14**, but there exists the stacking involving thiophene rings, as shown in Fig. 10. In the structure of **15**, the overlap of Pc moieties is only marginal type K Pc π–π interactions. Here, stacks of Pcs are separated by double layers of axial substituents. In both **13** and **14** the Pc-planes of molecules belonging to different stacks and are not parallel but contact in a herringbone pattern, whereas in the triclinic crystal of **15**, all Pc moieties are parallel and the packing can be described as laminar. Very recently, Sosa-Sánchez and co-workers [15] determined five single-crystal structures of Si^{IV}Pc[OOC(CH$_2$)$_n$CH$_3$]$_2$ with long alkyl carboxylates as axial ligands, where $n = 7$ (**16**), 10 (**17**), 12 (**18**), 13 (**19**) and 20 (**20**). The coordination surroundings of Si^{IV} ions in these complexes are similar to those of complexes **13**–**15**. There exist types K and M Pc π–π overlaps for complex **16**; J-type overlap for **17, 18** and **20**; and L-type overlap for **19**, giving a 2D supramolecular structure for **16** and 1D structures for **17–20**.

3.2.3 Unsymmetrical Hexacoordinated Phthalocyanines

Compared with symmetrical hexacoordinated MPcs, unsymmetrical hexacoordinated MPcs are relative limited due to the difficulties of synthesis. In the last 5 years, only a few such cases have been published. In 2002, Sosa-Sánchez [36] reported a chiral Pc $Si^{IV}Pc(naproxene)_2$ (**21**). The molecular structure determination shows that the complex crystallizes in a non-centrosymmetric space group due to the inherent chirality of the naproxene ligands. The molecules of complex **21** form 1D polymers through weak type K Pc π–π interactions. Recently, Bryce and Beeby [37] unexpectedly obtained one complex β-$Si^{IV}Pc(3-TA)_2$ (**22**) that crystallizes in the space group C_2/c. In α-$Si^{IV}Pc(3-TA)_2$ (**14**) the thienyl rings are stacked to the Pc moiety. However, in **22** one thienyl ring is near-parallel and the other nearly perpendicular to the Pc moiety. The molecule stacking in complex **22** is also different from that in **14**. In **22**, the β-$Si^{IV}Pc(3-TA)_2$ molecules only form dimers via type C Pc π–π overlap. Recently, Galezowski and Ercolani [38] reported unsymmetrical complexes $Co^{III}Pc(CH_3)(Py)$ (**23**) and $[Cr^{III}Pc(N_3)(OH)Mn^{III}TPP] \cdot 2ClNP$ (**24**) with different axial ligands. The molecules of complex **23** form a 1D ladder-like supramolecular structure by the combination of types J and M Pc π–π overlaps. The Co···Co separation in complex **23** is 8.917 Å. For complex **24**, the molecules only form dimers through the type C Pc π–π interactions due to steric effects of the large ligand TPP and solvate molecules ClNP. The intradimer Cr···Cr separation is 7.789 Å in **24**.

3.3 Neutral Parent Out-of-Plane Phthalocyanines

If the central ion is too large to be able enter the inner cavity of the Pc macrocycle or there is a strong unsymmetrical axial ligand, the central metal ion rests "atop" of the Pc ring (see Fig. 11, and Table 2 for details). The degree of shift of the metal ion from the Pc ring can be calculated from the deviation of the metal ion from the $(N_{iso})_4$ mean plane and the corresponding N_{iso}–M–N_{iso} angle (ϕ). In this case, the Pc macrocycle adapts to this special bonding situation by deformation,

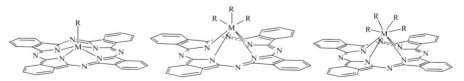

Fig. 11 Schematic structures of neutral parent out-of-plane Pcs

New Progress in Monomeric Phthalocyanine Chemistry

Table 2 Neutral parent out-of-plane Pcs

Compound	Metal dev.[a] Å	M$\cdots$M[b] Å	π–π verlap[c]	Structure[d]	Ref.
[Be^{II}Pc(H$_2$O)]·EE	0.264	5.271	A, B	1D	[39]
[Zn^{II}Pc(H$_2$O)]·2DMF	0.380	5.824	F, G	2D	[40]
Mg^{II}Pc(H$_2$O)	0.454	6.607	B, I	1D	[41]
Mg^{II}Pc(H$_2$O)·Py	0.444	4.535	A, B	1D	[42]
Mg^{II}Pc(H$_2$O)·2(3-ClPy)	0.496	6.710	F	Dimer	[43]
Mg^{II}Pc(H$_2$O)·0.5MEA	0.488	5.922	F	Dimer	[44]
Mg^{II}Pc(H$_2$O)·MEA	0.491	5.788	B, F	1D	[44]
Mg^{II}Pc(H$_2$O)·1.5MEA	0.492	4.576	E, F	Dimer	[44]
Mg^{II}Pc(CH$_3$OH)	0.437	4.602	A, B	1D	[45]
Mg^{II}Pc(EE)	0.458	4.615	A, B	1D	[46]
Hf^{IV}Pc(TDOFe)·3C$_6$H$_6$	1.104	7.720	F	Dimer	[47]
Sn^{IV}Pc[OOC(CH$_2$)$_4$CH$_3$]$_2$	0.966	6.917	F, L	1D	[16]
Sn^{IV}Pc[OOC(CH$_2$)$_4$CH$_3$]$_2$·CHCl$_3$	0.935	7.351	F, J	1D	[16]
Sn^{IV}Pc[OOC(CH$_2$)$_6$CH$_3$]$_2$	0.944	6.989	F, L	2D	[16]
Sn^{IV}Pc[OOC(CH$_2$)$_8$CH$_3$]$_2$	0.961	6.350	F, L	1D	[16]
Sn^{IV}Pc[OOC(CH$_2$)$_8$CH$_3$]$_2$	0.956	7.205	D, F, L	2D	[48]
Sn^{IV}Pc[OOC(CH$_2$)$_{12}$CH$_3$]$_2$	0.957	7.948	I, L	1D	[48]
Sn^{IV}Pc[OOC(CH$_2$)$_{14}$CH$_3$]$_2$	0.944	7.545	I, J, L	2D	[16]

[a,b,c,d] As for Table 1

which can be estimated by the dihedral angle between the opposite isoindole units. The MPc skeletons in these complexes usually exhibit dome- or cap-shaped distortion. Attributed to the absence of axial ligands at one side, these complexes usually form concave–concave overlapping dimers via very strong π–π interactions. If the steric effects of axial ligands at another side are suitable, these dimers can further slipped-stack to form convex–convex overlapping 1D or 2D supramolecular structures through relatively weaker Pc π–π interactions.

3.3.1 Pentacoordinated Pcs

Pentacoordinated metal Pc with bent macrocycle is relatively prevalent in the family of Pc. In 2006, Kubiak and co-workers [39] examined the single-crystal structure of complex Be^{II}Pc(H$_2$O)·EE (**25**). The coordination polyhedron of the metal atom, consisting of four isoindole N atoms and the apical water molecule, is a square pyramid. The Be^{II} atom is shifted out of the (N$_{iso}$)$_4$ mean plane by 0.264(6) Å. However, it should be expected that the cavity formed by four N$_{iso}$ atoms is very spacious for the small Be^{II} ion. The reason for the significant shift of the Be^{II} ion from the Pc centre might be due to the coordinated water molecule, which plays an essential role as a link between the Be^{II}Pc and ethoxyethanol units for the stabilization and arrangement of the structural units in the crystal. The molecules of complex **25** form 1D polymer via types A and B Pc π–π overlaps and O–H$\cdots$N$_{aza}$ hydrogen bonds (see Fig. 12). The shortest Be$\cdots$Be distance in the polymer is 5.271 Å. Very

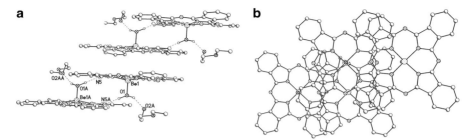

Fig. 12 Parallel **a** and perpendicular **b** views of the 1D Pc chain arrangement in **25**

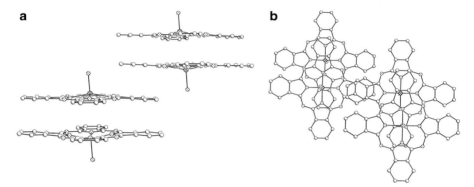

Fig. 13 Parallel **a** and perpendicular **b** views of the 2D Pc sheet arrangement in **26**; solvate DMF molecules are omitted for clarity

recently, the crystal structure of $[Zn^{II}Pc(H_2O)] \cdot 2DMF$ (**26**) has been solved by Fu and co-workers [40]. The results indicate that the Zn^{II} ion in this complex is also pentacoordinated. Notably, the Zn^{II} atom is at the inversion centre, and the distance between the Zn^{II} atom and the least square plane defined by $(N_{iso})_4$ toward the water molecule is about 0.380 Å. There are strong Pc π–π interactions (type F and G) in complex **26**, leading to a 2D supramolecular layer, as shown in Fig. 13. The shortest intermolecular $Zn \cdots Zn$ separation in this complex is 5.824 Å.

$Mg^{II}Pcs$ are the most widely investigated pentacoordinated parent Pc complexes. In the past 5 years, eight such compounds (**27–34**) with water and alcohols as axial ligands have been published [41–46]. The solid-state arrangements of these complexes display very different stacking patterns with the change of axial ligands and of solvate molecules. For complex **27**, there are two independent $Mg^{II}Pc(H_2O)$ molecules in the asymmetric unit with very similar geometries [41]. In both molecules, the Mg^{II} ions are significantly displaced from the $(N_{iso})_4$ mean plane towards the oxygen atom from the water molecule, with an average displacement of 0.454(3) Å. The supramolecular structure is built up from two crystallographically different $[Mg^{II}Pc(H_2O)]_2$ dimers (see Fig. 14). The dimers are stacked in a herringbone fashion along the b-axis, giving 1D supramolecular columns. Besides the strong type I Pc π–π interactions in the dimers, the type B Pc π–π interaction

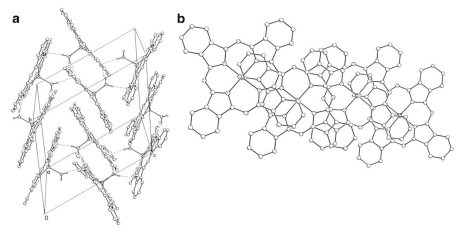

Fig. 14 Molecular arrangement in the crystal packing **a** and single 1D columnar structure **b** in **27**

between dimers along the columns is also very important. The columns then are linked by O–H···N$_{aza}$ hydrogen bonds into a 2D sheet-like supramolecular structure. For complex Mg^{II}Pc(H$_2$O)·Py (**28**), the molecules first form dimers via type A Pc π–π interaction, then these dimers stack one to another by type B Pc π–π interaction, giving rise to 1D supramolecular chains [42]. Unlike complex **27**, the chains based on the Pc π–π interactions in **28** are parallel to each other. For complexes **29**, **30** and **32**, the molecules only form dimers through the Pc ring π–π interactions [43–45]. For **31**, **33** and **34**, the Mg^{II}Pc moieties form 1D chains by the combination of two types of Pc plane π–π interactions (type B and F for **31**, and type A and B for **33** and **34**) [44–46]. The separation of Mg^{II} ion from the mean (N$_{iso}$)$_4$ plane ranges from 0.437 to 0.496 Å, which are slightly larger than those of complexes **25** and **26** for Zn^{II} and Be^{II} ions, respectively. The very different arrangements of molecules in complexes **27–34** further indicates that the axial ligands and the solvate molecules have significant influence on the supramolecular assemblies of MPcs.

3.3.2 Heptacoordinated Pcs

Heptacoordinated neutral parent Pcs with crystal structures are very scarcely described in the literature of the past 5 years. In 2005, Voloshin's group [47] reported a heptacoordinated complex Hf^{IV}Pc(TDOFe)·3C$_6$H$_6$ (**35**) with a tridentate tridioximates macrobicyclic iron^{II} complex as axial additional ligand. An apparent type F Pc π-stacking between the "base-to-base" oriented neighbouring molecules is a characteristic and specific feature of the crystal structure of **35**. The π-systems of two coplanar "overlapping" pyrrol fragments of Pc macrocycles of these molecules interact at a distance of 3.350 Å. The four coordinated nitrogen (N$_{iso}$) atoms of these macrocycles are located in one plane, whereas the coordinating Hf^{IV} ion is displaced by 1.105(1) Å from the (N$_{iso}$)$_4$ plane. The intradimeric Hf···Hf distance is 7.720 Å.

3.3.3 Octacoordinated Pcs

Recently, investigations on octacoordinated neutral parent Pcs have mainly focused on Sn^{IV} Pc complexes with long alkyl carboxylates as two *cis* ligands. In 2004 and 2005, Beltrán and co-workers synthesized a series of such complexes, $Sn^{IV}Pc[OOC(CH_2)_nCH_3]_2$, based on *trans*-coordinated $Sn^{IV}Pc(Cl)_2$ precursor and long alkyl acids, where $n = 4$ (**36** and **37**), 6 (**38**), 8 (**39** and **40**), 12 (**41**) and 14 (**42**) [16, 48]. The X-ray structures of complexes **36–42** show that the chlorine–carboxylate ligand exchange during the course of the reaction results in an overall change in configuration from *trans* to *cis*. The coordination geometry of the tin atom is approximately SnN_4O_4 square antiprismatic. The presence of the $(RCOO)_2Sn^{IV}$ moiety induces a deformation of the Pc(2-) ligand, which is directly responsible for the elongation of the N–Sn bonds. This, in turn, allows the coordination number to increase, yielding the hypercoordinated compounds with a nanocap shape. The displacements of the Sn^{IV} ions from the $(N_{iso})_4$ core are within the narrow range 0.935–0.966 Å. The intermolecular $Sn \cdots Sn$ distance in **36–42** ranges from 6.350 to 7.948 Å. The results confirmed that the orientation and length of hydrocarbon chains in the alkyl carboxylates have important effects on the molecular arrangements in these complexes (Table 3). It is noteworthy that the hydrocarbon chains in the myristic acid derivative **41** have a parallel arrangement, which is situated between two columns linked by type I and L Pc π–π interactions. Interestingly, there are three different Pc π–π overlaps, giving 2D sheet-like supramolecular structures for complexes **40** (type D, F and L) and **42** (type I, J and L, as shown in Fig. 15).

3.4 Peripherally Symmetrical Substituted Phthalocyanines

As discussed above, the introduction of axial ligands indeed effectively decreases the Pc π–π interactions. In this section, we will recommend the crystal structures of

Table 3 Peripherally symmetrically tetrasubstituted Pcs

Compound	Metal dev.[a] Å	$M \cdots M^b$ Å	π–π Overlap[c]	Structure[d]	Ref.
$H_2Pc(DMPO)_4 \cdot ox$	–	–	–	–	[49]
$Pd^{II}Pc(DMPO)_4 \cdot ox$	0	8.786	–	–	[17]
$Co^{II}Pc(DMPO)_4(H_2O)$	0	8.576	–	–	[17]
$Zn^{II}Pc(DMPO)_4(H_2O)$	0	8.619	–	–	[17]
$Mn^{III}Pc(DMPO)_4(Cl) \cdot H_2O \cdot THF$	0.290	10.436	–	–	[17]
$Co^{II}Pc(TMPO)_4 \cdot 2CH_3CH_2OH$	0	8.056	–	–	[50]
$Pb^{II}Pc(PO)_4$	1.304	5.986	B	–	[51]
$Zn^{II}Pc(PO)_4(H_2O)$	0.346	6.425	B	Dimer	[52]

[a,b,c,d]As for Table 1

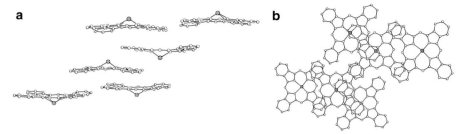

Fig. 15 Parallel **a** and perpendicular **b** views of the 2D Pc sheet in **42**; axial ligands are omitted for clarity

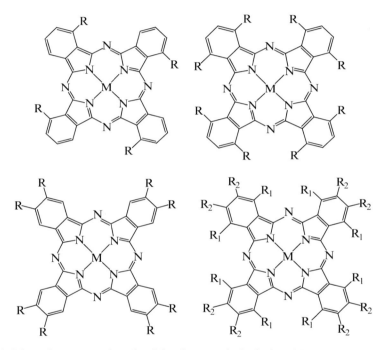

Fig. 16 Schematic representation of peripheral symmetrical substituted Pcs

peripherally substituted Pcs including symmetrical tetra-, octa- and hexadecasubstituted Pcs (see Fig. 16), and unsymmetrical AAAB, AABB, ABAB and ABBB Pcs either with or without axis ligands.

3.4.1 Tetrasubstituted Phthalocyanines

Symmetrical tetrasubstituted Pcs recently characterized by X-ray single diffraction techniques are mainly the 1,8,15,22-tetrasubstituted Pcs, which arise from

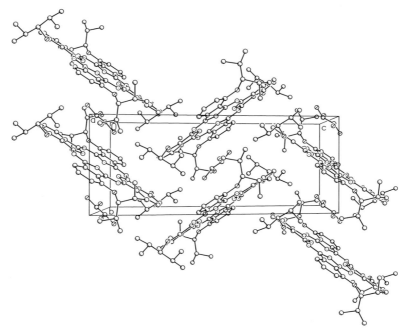

Fig. 17 Molecular arrangement of **43** in the unit cell

3-substituted phthalonitriles (see Table 3). In 2002, Ng and co-workers [49] synthesized a metal-free Pc H$_2$Pc(DMPO)$_4$ · ox (**43**). The results indicate that the molecular structure contains an inversion centre relating the two halves of the molecule, and the packing of molecules in the lattice is shown in Fig. 17. The Pc rings are stacked in a herringbone fashion along the crystallographic *b*-axis with a stacking angle of 55.0° and an interplanar distance of 7.22 Å. This arrangement is similar to that of the unsubstituted Pcs (both α and β forms), but the interplanar distance is much larger for **43**, probably due to the bulky DMPO substituents acting as the spacers. It is noteworthy that oxalic acid molecules intercalate between the Pc rings in this compound and are not hydrogen-bonded in any conventional or unconventional manner.

Two years later, the same group [17] reported the crystal structures of four metal complexes Pd^{II}Pc(DMPO)$_4$·ox (**44**), Zn^{II}Pc(DMPO)$_4$(H$_2$O) (**45**), Co^{II}Pc(DMPO)$_4$(H$_2$O) (**46**) and Mn^{III}Pc(DMPO)$_4$(Cl)·H$_2$O · **THF** (**47**), still based on bulky DMPO-substituted phthalonitriles. Complex **44** is a 1:1 inclusion complex of Pd^{II}Pc(DMPO)$_4$ and C$_2$H$_2$O$_4$. Similar to compound **43**, the guest species are not hydrogen-bonded and simply hang between the Pc rings in **44**. Complexes **45** and **46** both co-crystallize with a water molecule, which occupies a site with half occupancy, giving square pyramidal coordination geometries. The metal ions in complexes **44**–**46** lie strictly in the (N$_{iso}$)$_4$ plane. The Pc rings in complexes **44**–**46** are arranged in a herringbone manner along the crystallographic *b*-axis with a large

New Progress in Monomeric Phthalocyanine Chemistry 141

interplanar distance (6.98–7.40 Å), which is very similar to that of the metal-free analogue **43**. In complex **47**, the Mn^{II} centre is coordinated by four N_{iso} atoms of the Pc ring and one terminal chloro group, forming a slightly distorted square pyramid. In contrast to the zinc and cobalt analogues, the Mn^{III} ion in **47** is displaced 0.291 Å above the $(N_{iso})_4$ plane towards the apical chloro ligand. Unlike neutral out-of-plane Pcs with axial ligands, which always form dimers via strong Pc π–π overlaps, no such π–π interactions are found in **47** due to the steric effects of the bulky substitutes. In 2004, Huang and co-workers [50] determined the crystal structure of complex $Co^{II}Pc(TMPO)_4 \cdot 2CH_3CH_2OH$ (**48**), which has a similar molecular arrangement to that of complexes **43–46**.

In 2004, Jiang and co-workers [51] reported a 1,8,12,25-tetrasubstituted Pc complex $Pb^{II}Pc(PO)_4$ (**49**). The results indicate that this compound, having a non-planar structure, crystallizes in the monoclinic system with a $P2_1/c$ space group. Each unit cell contains two dimers of enantiomeric molecules, which are linked by weak coordination of the Pb atom of one molecule with an N_{aza} atom and its neighbouring oxygen atom from the alkoxy substituent of another molecule, forming a pseudo-double-decker supramolecular structure in the crystals with a short ring-to-ring separation of 2.726 Å. The coordination polyhedron of the lead is thus essentially a slightly distorted trigonal prism. However, due to the larger ionic size, the divalent lead ion cannot situate in the central hole of $Pc(PO)_4$ but sits atop, 1.305 Å above the $(N_{iso})_4$ mean plane. A strong type B Pc π–π interaction is observed in the dimers. The presence of peripherally substituted PO groups prevents these dimers from further stacking via Pc π–π interactions.

Very recently, the same group [52] synthesized a 1,8,12,25-tetrasubstituted Pc complex $Zn^{II}Pc(PO)_4(H_2O)$ (**50**). The compound crystallizes in the monoclinic system with a $P2_1/c$ space group with two pairs of enantiomeric molecules in a unit cell like complex **49**. In each monomeric $Zn^{II}Pc(PO)_4(H_2O)$ unit, the zinc ion is coordinated with four isoindole nitrogen atoms and one oxygen atom from the Pc ligand and an axial coordinated water molecule, giving a slightly distorted square pyramid geometry. Similar to other pentacoordinated $M^{II}Pcs$, the divalent zinc ion in this complex does not situate in the central hole of $Pc(PO)_4$ but sits atop, 0.346 Å above the $(N_{iso})_4$ plane. As a result, the substituted $Pc(PO)_4$ ring adopts a conformation that is domed towards the zinc cation. It is noteworthy that the two Pc molecules are bound to each other via four hydrogen bonds, leading to the formation of a pseudo-double-decker supramolecular structure $[Zn^{II}Pc(PO)_4(H_2O)]_2$. A relative strong type B Pc π–π interaction is also observed in the dimers, and the peripherally substituted PO groups hinder these dimers' further stacking via Pc π–π interactions, which are similar to those observed in complex **49**.

3.4.2 Octasubstituted Phthalocyanines

In the past 5 years, about ten 1,4,8,11,15,18,22,25-octasubstituted Pcs have been structurally characterized, as listed in Table 4. One typical example of this type of Pc is the saddle-shaped distortion of Pc rings, and the extent of distortion can be

142   Z. Ni et al.

Table 4 Peripherally symmetrically octasubstituted Pcs

Compound	Metal dev.[a] Å	M$\cdots$M[b] Å	π–π Overlap[c]	Structure[d]	Ref.
H$_2$Pc(nBuO)$_8$	–	–	I, L	1D	[53]
H$_2$Pc(nAmO)$_8$	–	–	K	1D	[54]
Cu^{II}Pc(EO)$_8$	0.022	5.040	F, L	1D	[55]
Ni^{II}Pc(nBuO)$_8$	0.001	7.584	G, L	2D	[56]
Cu^{II}Pc(iPO)$_8$	0.041	6.180	I	Dimer	[57]
Co^{II}Pc(iPO)$_8$	0.034	6.386	I	Dimer	[57]
Ni^{II}Pc(iPO)$_8$	0.021	6.194	I	Dimer	[57]
Pb^{II}Pc(HT)$_8$	1.323	5.213	B, J, K	1D	[58]
Ni^{II}Pc(hexyl)$_8$	0	5.850	I	1D	[59]
In^{III}Pc(hexyl)$_8$(Cl)	0.703	7.729	I, L	1D	[60]
In^{III}Pc(hexyl)$_8$(4-FP)	0.833	8.942	G, I	1D	[60]
Zn^{II}Pc(Ph)$_8$(Py)$\cdot$2Py$\cdot$Ph	0.398	8.248	–	–	[61]
Zn^{II}Pc(DiPPO)$_8$(H$_2$O)$\cdot$H$_2$O	0.419	13.677	–	–	[18]
H$_2$Pc(DHCP)$_4$	–	–	F, I	1D	[62]
Ru^{II}Pc(15-C-5)$_4$(TED)$_2\cdot$7CHCl$_3$	0	15.931	–	–	[63]

[a,b,c,d] As for Table 1

estimated by the dihedral angles between opposite indole rings. Alkyloxy groups
are the most common substitutes in this type of Pc due to the easier preparation of their precursors. In 2002 and 2003, Huang [53] and Ercolani [54] reported
octabutyloxy- and octaamyloxy-substituted metal-free Pcs H$_2$Pc(nBuO)$_8$ (**51**) and
H$_2$Pc(nAmO)$_8$ (**52**), respectively. The Pc skeleton of **51** is a highly saddle-shaped
distortion, whereas the distortion of complex **52** is relative small; the reason might
be attributed to the different steric orientation of alkoxy groups in the two compounds. The two compounds both form 1D supramolecular structures via Pc π–π
interactions. In 2005, Wang and co-workers [55] examined the crystal structure
of Cu^{II}Pc(EO)$_8$ (**53**). The molecules still form a 1D supramolecular structure via
type F and L Pc π–π interactions. In the same year, Rodgers [56] synthesized a
octabutyloxy-substituted compound Ni^{II}Pc(BuO)$_8$ (**54**). The structural data reveal
that the macrocycle assumes a saddle conformation, with the indole rings tilted
alternately up and down, almost as rigid bodies. The opposite indole rings form
dihedral angles of 31.5(1)$^\circ$ and 32.4(1)$^\circ$, respectively. In the crystal packing, the
molecules form a 2D sheet-like supramolecular structure through type F and L overlaps. Very recently, Lin [57] reported a series of *iso*-pentoxy-substituted complexes
M^{II}Pc(iPO)$_8$ (Cu (**55**), Co (**56**) and Ni (**57**)) with crystal structures. The X-ray analysis reveals that molecules aggregate to dimers via strong type I π–π overlap in the
crystals of the three complexes. The steric congestion of the substituents spread-
ing outwards holds back molecules of one dimer from aggregating with that of the
neighbouring dimer.

Similar to the above-mentioned alkoxy, alkylthio groups are also excellent peripheral substitutes. In 2003, Cook [58] determined the crystal structure of substituted Pc complexes Pb^{II}Pc(HT)$_8$ (**58**) with eight hexylthio groups attached to the

α-sites of the Pc macrocycle. There are two dependent $Pb^{II}[Pc(HT)_8]$ units in the crystal structure. The two lead atoms are each tetracoordinated, and lie 1.339(4) and 1.306(4) Å out of the square plane of the four N_{iso} atoms. Similar to complex **49**, the Pb^{II} ions in this complex exhibit essentially slightly distorted PbN_5S_2 trigonal prisms. The arrangements of molecules form 1D chain-like supramolecular structures via type B and K Pc π–π interactions for one chain and type B and J Pc π–π interactions for another chain. The two neighbouring chains are parallel to each other in the crystal.

Long flexible alkyl groups have been exploited widely as peripheral substitutes. In 2003, Helliwell and co-workers [59] investigated the temperature-resolved structural behaviour of $Ni^{II}Pc(hexyl)_8$ (**59**). The molecules of this complex at 293 K stack one to another via type G Pc π–π overlap, giving a 1D chain-like supramolecular structure. In 2005, Cook's group [60] reported two symmetrical octasubstituted complexes $In^{III}Pc(hexyl)_8(Cl)$ (**60**) and $In^{III}Pc(hexyl)_8(4\text{-FP})$ (**61**) with bent Pc macrocycles. The indium atoms in the two complexes are displaced from the central cavities of the macrocycles, lying 0.703 and 0.833 Å above the $(N_{iso})_4$ mean plane for **60** and **61**, respectively. Despite varying the axial ligand from a chloro group to the more space-demanding 4-fluorophenyl group, the two compounds form similar types of columnar stacks through Pc π–π overlaps. For **60**, the molecules of one column are tilted with respect to those in an adjacent column and thus exhibit a type of herringbone arrangement, whereas the molecules of complex **61** pack with their central planes parallel.

Recently, a rigid phenyl group has been employed as peripheral substituted ligand. In 2005, Kobayashi [61] synthesized a 1,4,8,11,15,18,22,25-octaphenyl-substituted Pc complex $Zn^{II}Pc(Ph)_8(Py) \cdot 2Py \cdot Ph$ (**62**) using H_2PcPh_8 as starting precursor. The crystal structure indicates that the molecular shape exhibits a highly deformed saddle-shaped structure with alternating up and down displacements of the isoindole units. The dihedral angles between the opposite indole rings are very large, amounting to $42.34°$ and $61.89°$, respectively. The Zn^{II} ion rests above the $(N_{iso})_4$ mean plane with a deviation of 0.398 Å. In this complex, the intermolecular Pc π–π interactions are very small due to the high distortion of Pc macrocycle, as shown in Fig. 18.

The reports on the crystal structure of 2,3,9,10,16,17,23,24-octasubstituted Pcs are relatively limited. To our best knowledge, there are only three such complexes that have been structurally characterized in the past 5 years [18, 62, 63]. Compared with the α-substituted Pcs, the peripheral β-site ligands have little influence on the deformation of Pc skeleton because the substitutes in this case are away from the Pc core. In addition, the ligands attached to the β-sites of Pcs can be either chain-like or cyclic groups. In 2005, McKeown [18] synthesized complex $Zn^{II}Pc(DiPPO)_8(H_2O) \cdot H_2O$ (**63**) using the large rigid $DiPPO$ group as peripheral substitute. The crystal structure is cubic and belongs to the exceptionally rare space group $Pn\text{–}3n$ with 12 Pc molecules in the unit cell. The Pc core of complex **63** is a shallow cone-shape with the central Zn^{II} ion and the oxygen atom of its axial ligand protruding from the molecular plane. As predicted, the $DiPPO$ substituents lie out of the plane of the Pc macrocycle and thereby prohibit the formation of columnar Pc

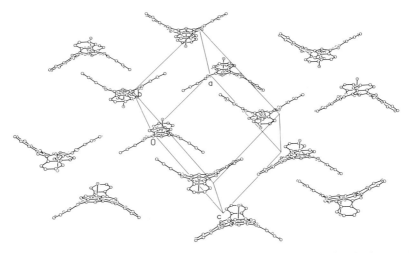

Fig. 18 Molecular arrangement in the crystal packing of **62**; all atoms not involved are omitted for clarity

stacks. In addition to the axial water ligand attached to the Zn^{II} centre of complex **61**, there are a further 24 water molecules per unit cell that appear to be associated through hydrogen-bonding interactions to the N_{aza} atoms of the Pc ring. Therefore, the complex is a Pc clathrate of cubic symmetry containing interconnected solvent-filled voids of nanometre dimensions. Very recently, the same group [62] reported the crystal structure of a novel metal-free Pc compound $H_2Pc(DHCP)_4$ (**64**) based on 2,2-dialkylindane precursor. The Pc macrocycle only shows slightly deformation. The molecules of complex **64** form a 1D columnar supramolecular structure via types F and I Pc π–π interactions.

In 2004, Nefedov and co-workers [63] reported a tetra-18-C-5-substituted complex $Ru^{II}Pc(DHCP)_4(TED)_2$ (**65**) with two axial TED ligands and seven solvate molecules $CHCl_3$. The Pc π–π interactions are completely prohibited by the combination roles of peripheral substituted and axial coordinated ligand as well as by solvate molecules.

3.4.3 Hexadecasubstituted Phthalocyanines

The hexadecasubstituted Pcs with crystal structures are shown in Table 5. In 2004, Kimura [64] prepared a metal-free Pc compound $H_2Pc(Et)_8(BzT)_8 \cdot 2CHCl_3$ (**66**), in which the α- and β-sites are occupied by eight ethyl and eight benzylthio groups, respectively. In the structure, two benzylthio groups on the same benzene ring are directed in opposite directions and are perpendicular to the plane of the Pc, while the two ethyl groups in close proximity are also oriented in alternate directions. It seems that the two sterically congested ethyl groups cause a slight distortion in the Pc skeleton. No obvious Pc π–π interactions are observed in this complex. Instead

Table 5 Peripherally symmetrically hexadecasubstituted Pcs

Compound	Metal dev.[a] Å	M···M[b] Å	π–π Overlap[c]	Structure[d]	Ref.
H$_2$Pc(Et)$_8$(BzT)$_8$ · 2CHCl$_3$	–	–	–	–	[64]
H$_2$PcF$_8$(PFiP)$_8$	–	–	A	Dimer	[65]
Zn^{II}PcF$_8$(PFiP)$_8$(ac)$_2$	0	12.106	–	–	[66]
Co^{II}PcF$_8$(PFiP)$_8$(ac)$_2$	0	11.981	–	–	[19]
Co^{II}PcF$_8$(PFiP)$_8$(TPIP)$_2$	0	13.477	–	–	[19]

[a,b,c,d] As for Table 1

Fig. 19 Overall molecular arrangement with fluorine-lined solid-state channels in complex **67**

strong π–π interactions between Pc and the phenyl ring from the benzylthio group are found, which link the molecules into a 1D supramolecular structure.

In 2002 and 2003, Diebold and Gorun [65] reported four perfluorinated metal-free Pcs and metal Pcs with eight fluorine atoms and eight perfluoroisopropyl appended to the α- and β-sites, respectively. Compound H$_2$PcF$_8$(PFiP)$_8$ (**67**) was the first perfluorinated metal-free Pc to be synthesized from the corresponding phthalonitrile. The crystal structure indicates that the Pc ring exhibits dome-like structural deformation in the solid state. The dihedral angle between the opposite isoindole units is ∼20°. Compound **67** forms slipped-stacked dimer via strong type A Pc π–π interaction due to the dome-like molecular shape, which decreases the intradimeric steric congestion. Interestingly, this complex displays very beautiful network molecular arrangements, as shown in Fig. 19. There are spacious fluorine-lined solid-state channels. The intermolecular interlocking of peripheral *iso*-perfluoroalkyl short-chains favours the formation of fluorine-lined solid-state channels while imposing a metal-induced-type molecular distortion. In 2002, they

[19, 66] reported three perfluorinated MPcs **68–70** with acetone or TPiP molecules as two axial ligands. The Pc rings in the three complexes are perfectly planar with the Co or Zn atoms at its geometric centre, which is very different from compound **66**. The CF$_3$ groups of the iC$_3$F$_7$ substituents direct above and below the phthalocyanine plane. Expectedly, no π–π stacking is observed in the solid state of the three complexes.

3.5 Peripherally Unsymmetrical Substituted Phthalocyanines

Reports of the crystal structures of peripherally unsymmetrical substituted Pcs are very scarce because of the difficulties of synthesis and separation (see Table 6 for details). Recently, Wang and co-workers [67] synthesized a *trans*-form α-disubstituted metal-free Pc compound H$_2$Pc(1,15-BTMPO) (**71**) by the "cross-condensation" method. The crystal structure reveals clearly that the distribution of the two alkoxy groups are arranged in a *trans* form. There are two independent molecules aggregated together to form a dimer in a unit cell, with a distance of 3.366 Å. The dimers are separated by the bulky alkoxy groups arranged at the four corners of the dimer.

In 2005, Kobayashi's group [61] synthesized a series of phenyl-substituted unsymmetrical MPcs, Zn^{II}Pc(Ph)$_2$(Py) · 2Py · MPh (**72**), Zn^{II}Pc(Ph)$_4$ · 3Py (**72**) and Zn^{II}Pc(Ph)$_6$ · 3Py (**73**). The coordination surroundings of Zn^{II} ions in the three complexes are the same as that of complex **59**. The crystal structures reveal that the overlap of the phenyl groups causes substantial deformation of the Pc ligands within the crystals, while strong π–π stacking in the remainder of the Pc moiety lacking phenyl substituents can suppress the impact of the deformation. The axial pyridine ligand prevents extensive π–π interactions on one side of the Pc and allows close π–π stacking (type E for **72** and **73**, type F for **74**) at the opposite side, giving a dimeric structure (see Fig. 20 for complex **74**). In the same year, Cook [60] reported a unsymmetrical substituted complex In^{III}Pc(Br)(BuO)$_2$(hexyl)$_6$ (**75**). The molecular arrangement of this complex is similar to that of complex **59**.

In 2006, Liu and co-workers [68] reported a AAAB-type complex Zn^{II}Pc(DPP)$_2$ (tBuPO)$_6$(Py) (**76**) by a statistical condensation reaction between two different

Table 6 Peripherally unsymmetrical substituted phthalocyanines

Compound	Metal dev.[a] Å	M $\cdots$ M[b] Å	π–π Overlap[c]	Structure[d]	Ref.
H$_2$Pc(TMPO)$_2$	–	–	E	Dimer	[67]
Zn^{II}Pc(Ph)$_2$(Py) · 2Py · MPh	0.363	4.333	E	Dimer	[61]
Zn^{II}Pc(Ph)$_4$ · 3Py	0.381	4.477	E	Dimer	[61]
Zn^{II}Pc(Ph)$_6$ · 3Py	0.341	6.082	F	Dimer	[61]
In^{III}Pc(Br)(BuO)$_2$(hexyl)$_6$	0.710	8.015	I, L	1D	[60]
Zn^{II}Pc(DPP)$_2$(p-tBuPO)$_6$(Py)	0.388	5.458	I	Dimer	[68]

[a,b,c,d]As for Table 1

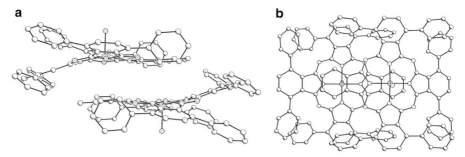

Fig. 20 Parallel **a** and perpendicular **b** views of the dimeric structure in **74**

phthalonitriles. The results indicate that the aromatic Pc skeleton is only distorted from perfect planarity. The Zn^{II} ion is situated out of the mean (N$_{iso}$)$_4$ plane by 0.388 Å. Two molecules of complex **76** form a concave–concave overlapping dimer through strong type I Pc π–π interactions. These dimers are prevented from further stacking through Pc π–π interactions by the axial and peripheral substituted ligands.

3.6 Ionic Phthalocyanines

Generally, the ionization of Pcs not only increases their solubility but also alters their π–π stacking structures, especially in the presence of large balanced ions. The Pc skeleton in these compounds can be integrally or partly oxidized cationic H$_2$Pc$^+$, or anionic Pc(1-) and Pc(1-). According to the charge type of Pc moiety, ionic Pcs can be classified as cationic Pcs and anionic Pcs. The cationic and anionic Pcs published recently are listed in Tables 7 and 8, respectively.

3.6.1 Cationic Phthalocyanines

In 2002, Ibers and co-workers [69] synthesized two integrally oxidized metal-free Pc compounds [H$_2$pc][IBr$_2$] (**77**) and [H$_2$pc]$_2$[IBr$_2$]Br · C$_{10}$H$_7$Br (**78**) by chemical oxidation. The extended structure of **76** comprises slipped Pc columns via type E and F Pc π–π interactions. The Pc rings stack along the a-axis in adjacent columns at 70° to one another. IBr$_2^-$ ions occupy the interstitial columns. Complex **77** forms slant stacks (types A and B) of Pc rings with IBr$_2^-$ ions, Br$^-$ ions, and 1-bromonaphthalene molecules in the adjacent, parallel columns. In the same year, Ibers and co-workers [70] reported a series of partially oxidized Pc compounds **79–88** as listed in Table 8. Complexes **79–84** are essentially isostructural. The molecular structures of the seven compounds are all comprised of a stack of three MPc trimers with inter-ring distance of about 3.16 Å. Unlike the above-discussed slipped-stacks of Pc rings, the centres of the three MPc units in a trimer are almost linear. The inner and outer Pc planes of the trimer are staggered by

Table 7 Cationic phthalocyanines

Compound	Metal dev.[a] Å	$M\cdots M^{b}$ Å	$\pi-\pi$ Overlap[c]	Structure[d]	Ref.
$[H_2Pc][IBr_2]$	–	–	E, F	1D	[69]
$[H_2Pc]_2[IBr_2]Br\cdot C_{10}H_7Br$	–	–	A, B	1D	[69]
$[H_2Pc]_3[AsF_6]_2\cdot C_{10}H_7Cl$	–	–	A, N	1D	[70]
$[H_2Pc]_3[SbF_6]_2\cdot C_{10}H_7Cl$	–	–	A, N	1D	[70]
$[Ni^{II}Pc]_3[SbF_6]_2\cdot C_{10}H_7Cl$	0.023	3.159	A, N	1D	[70]
$[Cu^{II}Pc]_3[SbF_6]_2\cdot C_{10}H_7Cl$	0.011	3.176	A, N	1D	[70]
$[Cu^{II}Pc]_3[AsF_6]_2\cdot C_{10}H_7Cl$	0.014	3.185	A, N	1D	[70]
$[Ni^{II}Pc]_3[ReO_4]_2\cdot C_{10}H_7Cl$	0.003	3.215	A, N	1D	[70]
$[Cu^{II}Pc]_3(ReO_4)_2$	0.123	3.160	L, N	1D	[71]
$[Sb^{III}Pc]_4[Sb_6I_{22}]$	0.991	5.760	E	Dimer	[72]

[a,b,c,d] As for Table 1

Table 8 Anionic phthalocyanines

Compound	Metal dev.[a] Å	$M\cdots M^{b}$ Å	$\pi-\pi$ Overlap[c]	Structure[d]	Ref.
$[PXX][Fe^{III}Pc(CN)_2]$	0.012	7.680	C and G	1D	[73]
$[PXX]_2[Co^{III}Pc(CN)_2]$	0	7.295	C and G	2D	[74]
$[PXX]_2[Co^{III}Pc(CN)_2]\cdot CH_3CN$	0.004	7.825	C and G	2D	[75]
$[PXX]_4[Co^{III}Pc(CN)_2]\cdot CH_3CN$	0	10.406	G	2D	[75]
$TPP[Co^{III}Pc(Cl)_2]_2$	0	7.537	C	1D	[76]
$TPP[Co^{III}Pc(Br)_2]_2$	0	7.609	C	1D	[76]
$TPP[Mn^{III}Pc(CN)_2]\cdot CH_2Cl_2$	0	8.986	J, K	1D	[20]
$(nBu_4N)[Zr^{IV}Pc(OPh)_3]\cdot Et_2O$	1.291	12.132	–	–	[77]
$(nBu_4N)[Hf^{IV}Pc(OPh)_3]\cdot Et_2O$	1.267	11.904	–	–	[77]
$(nBu_4N)_2[Zr^{IV}Pc(tcc)_2]$	1.277	13.157	–	–	[77]
$(Et_4N)_2[Zr^{IV}Pc(O_2CO)_2]$	1.225	8.332	I	Dimer	[77]

[a,b,c,d] As for Table 1

40–44° (see Fig. 21b) in the seven complexes. These trimers then form 1D columns (see Fig. 21a) through quasi type A and type L $\pi-\pi$ interactions for **79–84** and **85**, respectively.

In 2006, Janczak and Perpétuo [72] reported a cationic Pc complex $[Sb^{III}Pc]_4$ $[Sb_6I_{22}]$ (**86**) with a large $[Sb_6I_{22}]$ cluster as counterion. The two independent $Sb^{III}Pc$ segments in **86** belong to typical out-of-plane structures with Sb^{III} ions lying above the $(N_{iso})_4$ mean plane with an average 0.991 Å separation. In fact, the two Sb^{III} ions are weakly coordinated by the iodine atoms from the $[Sb_6I_{22}]$ cluster, giving hepta- and octacoordinated trigonal prism and square antiprism, respectively. The complex shows a very interesting supramolecular assembly, as shown in Fig. 22. Four cationic $[Sb(Pc)]^+$ units form one box-like substructure with a large balanced $[Sb_6I_{22}]$ cluster resided within it. Then, these box-like substructures slipped-stack along four directions via concave–concave overlapping of Pc macrocycles, forming a 2D supramolecular network structure.

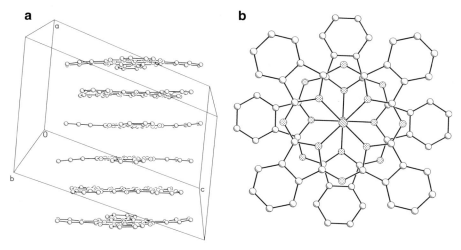

Fig. 21 Molecular arrangements in the crystal packing **a** and the Pc overlapping structure of adjacent Pc rings in trimer **b** of **79–84**

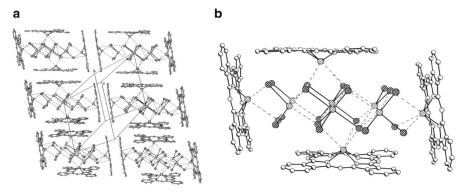

Fig. 22 Molecular arrangement in the crystal packing **a** and single box substructure **b** in **86**

3.6.2 Anionic Phthalocyanines

During the last 5 years, Inabe and co-workers [73–76] have used electrochemical oxidation to synthesize a series of partially oxidized ionic Pc compounds [PXX]$_n$[M^{III}Pc(CN)$_2$] (M = Fe or Co) (**87–90**) and [TPP][Co^{III}Pc(X)$_2$]$_2$ (X = Cl(**91**) or Br(**92**)) with two *trans* CN$^-$ groups, Cl$^-$ or Br$^-$ ions. The central metal ions all have distorted hexacoordinated octahedral geometries. The Pc skeletons in these complexes are nearly strictly planar and the central metal ions lie in the (N$_{iso}$)$_4$ planes. The M–C≡N angles are almost linear and the axial CN$^-$ groups, Cl$^-$ or Br$^-$ ions are relatively small. Therefore, these Pc complexes always form 1D column-like or 2D sheet-like Pc stacks via various Pc π–π interactions. In addition, the planar

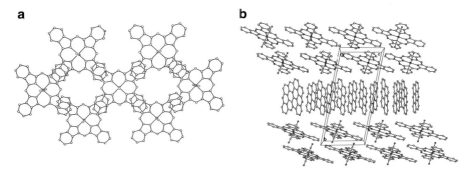

Fig. 23 Single sheet structure **a** and molecular arrangement in crystal packing **b** of **89**

PXX groups in complexes **87–90** can also slipped-stack to form columns located between the Pc columns and sheets. Since the crystal structures and supramolecular assembles of this type of phthalocyanine were well summarized in 2001 [12] and 2005 [79], we will present briefly only one typical Pc π–π supramolecular structure that has not been included in the previous section. As shown in Fig. 23, the Pc units in **89** form a 2D sheet along the *ac* plane. The overlap mode is type G Pc π–π interactions along both the *a*- and *c*-axes. This sheet further interacts with another sheet by type C overlap, forming a double sheet.

In 2005, Matsuda and co-workers [20] synthesized a $Mn^{III}Pc$ complex salt, $TPP[Mn^{III}Pc(CN)_2] \cdot DCM$ (**93**) with huge $[Mn^{III}Pc(CN)_2]^-$ anions and large TPP^+ cations. There are two crystallographically independent $[Mn^{III}Pc(CN)_2]^-$ molecular units. The central Mn^{III} ions of each unit lie at the inversion centre. The two unique $[Mn^{III}Pc(CN)_2]^-$ units form 1D supramolecular anionic chains via weak type J and K Pc π–π overlaps, and the adjacent chains built from different $[Mn^{III}Pc(CN)_2]^-$ unit are almost perpendicular to each other.

In 2002, Homborg's group [77] reported a series of anionic out-of-plane Pc complexes with large $[nBu_4N]^+$ and $[Et_4N]^+$ as balanced cations. The molecular structures and the coordination geometries of the central metal ions of anion segments in these complexes are very similar to those of the neutral out-of-plane Pcs (see Sect. 3.3). However, formation of the supramolecular structures based on the Pc π–π overlaps may take place due to the large balanced cations. For example, no significant Pc π–π stacks are observed in complexes $(nBu_4N)[M^{IV}Pc(OPh)_3] \cdot Et_2O$ (M = Zr (**94**) and Hf (**95**)) and $(nBu_4N)_2[ZrPc(tcc)_2]$ (**96**). Another common feature is that there exists an anionic layer formed by Pc anionic moieties and a cationic layer built from cationic $(nBu_4N)^+$ units, which may be responsible for the absence of Pc π–π overlapping in the three complexes. The molecular arrangement in crystal packing of **96** is shown in Fig. 24. Complex $(Et_4N)_2[Zr^{IV}Pc(O_2CO)_2]$ (**97**) forms concave–concave overlapping dimers via type I overlap, which is similar to that of common neutral out-of-plane Pcs. The difference in molecular arrangements between complexes **97** and **94–99** may be attributed to the fact that the $(Et_4N)^+$ cation is smaller than $(nBu_4N)^+$. In addition, the displacement of the central metal ions from the mean $(N_{iso})_4$ plane in the four complexes ranges from 1.225 to 1.291 Å.

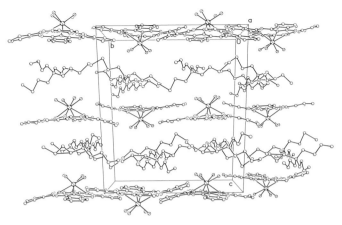

Fig. 24 Molecular arrangement in crystal packing of **96**

4 Properties of Phthalocyanines

Generally, the properties of monomeric Pcs are dependent on the nature of the central metal ions, on the position, steric effects and electronic properties of the axis and peripheral substituted ligands, on the extent of distortion of Pc skeletons, on the arrangements of molecules in the solid state or in solution, and on the degree of oxidation of Pc rings, as well as on some other external measuring conditions. In this section, we will discuss the UV–vis spectroscopic, conducting, magnetic and catalytic properties of some Pcs with crystal structures.

4.1 UV–vis Spectra

In general, the representative MPc has four main UV–vis absorption bands, and the spectrum is dominated by the intense Q and Soret (B) bands. A single intense Q-band absorption is usually observed in the visible region (620–800 nm) and is assigned to a $\pi \rightarrow \pi^*$ doubly degenerated transition (a_{1u}–e_g) within the macrocycle. However, unlike metal Pcs, metal-free Pcs usually have two bands of roughly equal intensity in the Q-band region due to the lower symmetry since the two inner hydrogen atoms cancel the degeneracy of molecular orbitals. The B band usually occurs in the near-ultraviolet region (300–350 nm) and is also attributed to a $\pi \rightarrow \pi^*$ transition (a_{2u}–e_g) of the macrocycle. Both Q and B bands are characteristic for the Pcs. Two relatively weak bands, as shoulders to the B and Q bands, often appear in the region 320–385 nm and 560–690 nm, respectively [4, 11].

4.1.1 Effects of the Central Metal Ions

The central metal ions can directly affect the electronic structures of Pc cores. Therefore, the UV–vis spectra of Pcs can change to some degree with the alternation of the metal ions. There are many results supporting this fact. The observed bands of a series of parent MPc with axial Py (**4–7**) or 4-MP ligands (**9**), and a family of alkoxyl-substituted MPcs (**44–47**) are listed in Tables 10 and 11. The Q-band position, however, is highly dependent on the metal centre in the order $Ru^{II} \ll Fe^{II} \ll Co^{II} < Mg^{II} \ll Mn^{II}$ for **4–7** and **9**, and $Pd^{II} < Co^{II} < Zn^{II} \ll Mn^{III}$ for complexes **44–47**.

4.1.2 Effects of Peripherally and Axially Substituted Ligands

Peripherally and axially substituted ligands also directly or indirectly tune the electronic absorption spectrum of Pc rings, the degree of which is mainly determined by the electronegativity and positions of substituents. Generally, axial ligands only have a small effect on the UV–vis absorption spectrum of corresponding Pcs, and the shifts of the Q and B bands are usually relatively small except the presence of very strong axial electron-withdrawing or electron-donating groups. The α-site substitutes usually have significant influence on the position of Pc absorption bands because they not only directly tune the electronic structure of Pc rings, but also result in the high distortion of Pc skeletons, as discussed above. Compared with the intense Q bands of parent unsubstituted Pcs observed usually in the 620–700 nm region (Table 9), the alkoxy, alkyl and alkylthio electron-donating groups at the α-sites of Pcs result in a bathochromic shift of the main absorption Q band from

Table 9 Some UV–vis absorption data of peripheral unsubstituted Pcs

Compound	Q				B	Ref.
in Py	676	644	609		350	[27]
in DMSO	672	641	604		345	[27]
in Py	692	660	620		360	[28]
in CH_2Cl_2	670		605		344	[25]
in CH_2Cl_2	650		594	415	322	[29]
in CH_2Cl_2	624				375	[31]
in $CHCl_2$	678	650	610		354	[34]
in $CHCl_3$	697	670	616		353	[15]
in $CHCl_3$	699	674	628		342	[15]
in $CHCl_3$	702	672	620		369	[15]
in $CHCl_3$	700	678	612		368	[15]
in $CHCl_3$	695	667	620		372	[15]
in EtOH	682		616		360	[16]
in EtOH	682		616		360	[16]
in EtOH	682		616		360	[16]
in EtOH	682		616		360	[48]
in EtOH	682		616		360	[16]

New Progress in Monomeric Phthalocyanine Chemistry

Table 10 Some UV–vis absorption data of peripheral α-substituted Pcs

Compound	Q				B	Ref.
in CHCl$_3$	694	664	623		314	[17]
in CHCl$_3$	702		631		313	[17]
in CHCl$_3$	711		638		322	[17]
in CHCl$_3$	775	693			337	[17]
in CHCl$_3$	748	670		444	333	[51]
in toluene	732	660		448	330	[56]
in THF	741					[57]
in THF	742					[57]
in THF	738					[57]
in THF	818	690		500	350	[58]
in THF	726					[60]
in THF	728					[60]

visible region to the near-IR region (700–820 nm) in solution (Table 10). In contrast to the α-site substituents, the influence of β-site substituents on the UV–vis absorption spectrum of Pc is obviously limited because they are further away from the Pc core than the α-site groups and have little effect on the distortion of the Pc core.

4.1.3 Effects of Distortion of Pc Skeleton

Deformation of the Pc skeleton also can directly affect the electronic structure of Pcs. Recently, the UV–vis spectroscopic properties of a series of deformed Zn^{II}Pcs were examined by Kobayashi and co-workers [66]. The results are shown in Table 11. All of the five complexes exhibit an intense single Q band in the region of 650–850 nm, and the absorption spectra show sizable red shifts of the Q band with increasing number of phenyl groups. Peripheral substitution usually results in an approximately linear shift of the Q-band energies as the number of substituents is increased. However, the results demonstrate that marked deviations from linearity are observed for the deformed complexes **73**, **74** and **62**. Analysis of the results of absorption spectra and electrochemical measurements reveals that a substantial portion of the red shift is attributed to the ring deformations. Molecular orbital calculations further support this conclusion. A moderately intense absorption band emerging at around 430 nm for highly deformed octaphenyl-substituted Zn^{II}Pc can be assigned to the HOMO $\rightarrow$ LUMO $+ 3$ transition, which is parity-forbidden for planar Pcs, but becomes allowed since the ring deformations remove the centre of symmetry.

4.2 Conductivity

Conductivity is one of the most important properties of Pcs, which has made them useful in many high-tech fields. In 2001, Inabe [12] summarized various electrically conducting neutral radical and partially oxidized salt crystals of $[Co^{III}Pc(CN)_2]^-$.

Table 11 UV–vis absorption data of $Zn^{II}Pc$ in pyridine with various phenyl groups

Compound	Q			B	Ref.
$Zn^{II}Pc$	674	609		345	[61]
	682	616		345	[61]
	704	636		343	[61]
	732	656	394	340	[61]
	786	697	429		[61]

Recently, the conducting properties of a series of fully and partially oxidized ion Pc complexes with single crystals have been investigated [73–76]. Herein we present recent progress on the conducting properties of Pcs from five aspects, which may offer new views for the designation of novel conducting Pc materials.

4.2.1 Effects of Arrangements of Pc Molecules

In 2004, Inabe and co-workers [74] compared the conductivity of complex **88** containing 2D Pc stacks with complexes $TPP[Co^{III}Pc(CN)_2]_2$ and $[PXX][Co^{III}Pc(CN)_2]$ having 1D chain-like and ladder-like supramolecular Pc stacks, respectively. The results indicate that the electrical resistivity in $[PXX]_2[Co^{III}Pc(CN)_2]$ is quite low ($10^{-3}\,\Omega\,cm$) and that the low resistivity is retained even at 5 K. Whereas the electrical resistivity of $TPP[Co^{III}Pc(CN)_2]_2$ and $[PXX][Co^{III}Pc(CN)_2]$ is higher than that of **88** ($10^{-2}\,\Omega\,cm$), the resistivity of these two complexes increases rapidly with lowering the temperature below 50 K and reaches about 10 $\Omega\,cm$ at 5 K. Since the HOMO–HOMO overlap integral value along the needle axis is nearly the same for the three crystals and the contribution from the partially oxidized 1D PXX columns is negligibly small, the difference in the transport properties of these complexes is attributed to the difference in the π–π stacking network structures. The temperature dependence of the thermoelectric power in **87** indicates clearly the metallic behaviour.

4.2.2 Effects of the Central Metal Ions

As discussed above, the nature of the central metal ions can directly affect the electronic structure of Pc rings. As a result, the nature of the central metal ions must be one important factor for determining the conductivity of Pcs. Inabe and co-workers [73] investigated the temperature dependence of the resistivity along the c-axis for complexes $PXX[M^{III}Pc(CN)_2]$ (M = Fe (**87**) or Co (**88**)), which are isomorphous and contain 1D two-leg ladder-like Pc stacks. The resistivity of the Fe salt ($3.3 \times 10^{-2}\,\Omega\,cm$) is nearly one order of magnitude higher than that of the Co salt ($6.0 \times 10^{-3}\,\Omega\,cm$). The Fe salt shows semiconducting behaviour in its electrical resistivity over the temperature range measured, while the isomorphous Co salt exhibits metallic behaviour above 100 K and very weak semiconducting behaviour

New Progress in Monomeric Phthalocyanine Chemistry 155

below 100 K. The difference in the transport properties between the two salts suggests that the conduction electrons in the Fe salt are seriously scattered by the local magnetic moment. The resistivity ratio of the Fe^{III} salt to the Co^{III} salt is about 10^5 at 25 K.

4.2.3 Effects of Axial Ligands

Recently, Inabe and co-workers [76] investigated the conductivity of three isomorphous complexes $TPP[Co^{III}Pc](L)_2$ (L = Cl (**91**), Br (**92**) or CN (**88**)) with different axial ligands. The temperature dependence of the resistivity for the three complexes indicates that the Br-ligated species has the highest electrical resistivity, and that the resistivity of CN-ligated complex is the lowest. As is well known, different axial ligands can result in different distances between the $Co^{III}Pc(L)_2$ moieties, thereby affecting the effectiveness of the π–π overlap. Therefore, the resistivity of the three complexes increases in the order $CN^- > Cl^- > Br^-$, which can be attributed to the influence of axial ligands on π–π overlap decreasing in the order $CN^- < Cl^- < Br^-$. This series of crystals exhibits an apparent semiconducting temperature dependence with a small activation energy ($E_A < 0.01\,eV$ for L = CN, $E_A = 0.008$–$0.015\,eV$ for L = Cl, and $E_A = 0.015$–$0.026\,eV$ for L = Br).

4.2.4 Effect of Pressure

The localized character of the charge carriers may be mobilized by applying pressure. Based on this consideration, Inabe and co-workers [75] investigated the conducting properties of Pcs under different pressures. The results indicate that pressure has a significant influence on the conductivity of **88**. The crystal shows semiconducting behaviour under ambient pressure. However, the conducting behaviour is metallic until 5 K under a pressure of 1.2 GPa.

4.2.5 Effects of Magnetic Field

In 2003, Inabe and co-workers [73] investigated the electrical conductivity under a magnetic field of 16 T for **87**. A magnetic field parallel to the c-axis (B//c) and perpendicular to the c-axis (B//a* and B//b) was applied. The resistivity of the Fe salt drastically decreases and a negative magnetoresistance is observed below 50 K for all directions, but the decrease in the resistivity is highly anisotropic to the field orientation. The B//a* magnetoresistance can be comparable to the B//b magnetoresistance and is much larger than the B//c magnetoresistance. The authors considered that the dependence of resistance on field orientation is highly consistent with the g-tensor anisotropy in the $[Fe^{III}Pc(CN)_2]$ unit, suggesting that the negative magnetoresistance originates from the large π–d interaction self-contained in the $[Fe^{III}Pc(CN)_2]$ unit.

In addition, Janczak [26] studied the conductivity property of complex **3** with a polycrystalline sample, and the results show that the conductivity is in the range $2.7 - 2.8 \times 10^{-2} \, \Omega^{-1} \, cm^{-1}$ at room temperature. Very weak temperature dependence of the conductivity and a metallic-like dependence in conductivity are observed in the range 300–15 K. Ibers and co-workers [70] investigated the electrical conductivity of partially oxidized complex **82** with a suitable single crystal and the results indicate its semiconductor nature ($E_A = 0.22 \, eV$).

4.3 Magnetic Properties

The investigations on the magnetic properties of monomeric Pc compounds mainly include the determination of the spin ground states of the central metal ions, magnetic coupling for metal–radical systems with fully or partially oxidized Pc rings, and the magnetic supra-exchange coupling through Pc $\pi-\pi$ interactions.

4.3.1 Spin States of the Central Metal Ions

Both high- and low-, and even intermediate-spin states can be found in Pc complexes, which have been the most appropriate medium for studying spin transition materials. In 2003, Janczak and co-workers [28] reported the magnetic properties of mononuclear complex **5** and its derivative $Mn^{II}Pc$. The results show the existence of overall antiferromagnetic interaction in **5** via a magnetic supra-exchange mechanism involving Pc $\pi-\pi$ overlap. Notably, at room temperature the effective magnetic moment calculated for **5** is $\mu_{eff} = 3.62 \mu_B$ (expected value $3.62 \mu_B$ for $S = 3/2$), indicating the intermediate-spin complex **5** with three unpaired electrons per molecule arising from the ground state configuration of $(a_{1g})^2(e_g)^2(b_{2g})^1$. The spin-only value for non-ligated complex **5** is also $S = 3/2$. However, the axial ligation of the $Mn^{II}Pc$ complex by pyridine changes its magnetic coupling between magnetic Mn^{II} centres, and ferromagnetic interactions are observed in the complex. The same group [25, 26] also investigated the spin ground states of complexes **6–8**. The results indicate that one unpaired electron localized on the d_z^2 orbital of Co^{II} ($e_g^4 b_{2g}^2 a_{1g}^1$, $S = 1/2$) ion in **6**, and that ligation of the intermediate-spin $Fe^{II}Pc$ by Py or 4-CP molecules leads to a change of the spin ground state configuration of the central ion from $S = 1$ ($Fe^{II}Pc$, $e_g^3 b_{2g}^2 a_{1g}^1$) to $S = 0$ (complexes **7** and **8**, $e_g^4 b_{2g}^2$).

In 2005, Matsuda and co-workers [20] investigated the magnetic properties of complex **93**. The magnetic susceptibility measurement reveals the Mn^{III} ion in the low-spin state ($d^4 \, S = 1$). However, the μ_{eff} at room temperature is $3.28 \mu_B$, which is considerably higher than the calculated spin-only value of $2.83 \mu_B$ for $S = 1$. The extraordinary μ_{eff} value should be due to the spin–orbit coupling effect.

New Progress in Monomeric Phthalocyanine Chemistry 157

4.3.2 Magnetic Coupling in Metal–Radical Systems

Compared with the sandwich Pc metal–radical systems, which exhibit excellent magnetic properties such as single-molecule magnet behaviours, reports on the magnetic properties of monomeric Pc complexes are relative limited. In 2003, Janczak and co-workers [26] investigated the magnetic properties of complex **3** with one-electron oxidized phthalocyaninato(1-) macrocyclic radical ligand. The results show that the effective magnetic moment at room temperature ($3.98\ \mu_B$) is slightly higher than the spin-only magnetic moment for Cr^{III}, $\mu_{spin-only} = 3.87\mu_B$ (d^3, $S = 3/2$), but it is considerably lower than that for the ferromagnetic coupling with the one-electron oxidized Pc(1-) radical ($S = 1/2$). The temperature dependence of the effective magnetic moment, μ_{eff}, shows the alternating ferro- and antiferromagnetic interactions in the system of the paramagnetic central Cr^{III} ion (d^3, $S = 3/2$) and surrounding π-conjugated radical ligand Pc(-1) ($S = 1/2$). The tendency of the magnetic moment in the range 2–300 K also indicates the quintet (300–25 K) and triplet (below 25 K) ground states, which might be produced by the ferro- and antiferromagnetic interactions between the paramagnetic Cr^{III} centres.

In the same year, Matsuda and co-workers [73] reported the magnetic properties of partially oxidized Pc complex **87** with one radical electron located partially distributing on both PXX and $[Fe^{III}Pc(CN)_2]_2$ components. Study of the magnetic susceptibility indicates the presence of overall antiferromagnetic interactions in the complex. The authors attributed the antiferromagnetic interactions to the antiferromagnetic coupling between $[Fe^{III}Pc(CN)_2]_2$ units via the π-electrons in the Pc rings. Moreover, spontaneous magnetization is observed below 8 K for the complex, which manifests weak ferromagnetism.

4.4 Catalytic Properties

The catalytic characteristics of Pcs have been widely investigated in the past few decades [11]. Herein, we present briefly the catalysis of a novel Pc complex, which was designed on the molecular level by Gorun and co-workers [19]. More interestingly, the crystal structures of both the catalyst itself and its catalytic intermediate have been determined. As discussed above, the introduction of bulky peripheral substituents encourages the formation of isolated monomeric species. If peripheral substituents are preferably perfluorinated, materials that are both soluble and resistant to self-oxidation might be yielded. Such materials may be suitable for homogeneous catalysis under harsh conditions, which include the presence of reactive species such as singlet oxygen and free radicals. Based on reasonable molecular design, they synthesized perfluorinated Pc complex **69** and investigated its catalytic characteristics. The results show that the enzyme-like complex can effectively couple phosphanes with acetone to produce ylide TPIP and water at ambient conditions, by using air as the sole reagent (Eq. 1):

$$Ph_3P + CH_3C(=O)CH_3 + \frac{1}{2}O_2 \rightarrow Ph_3P=CHC(=O)CH_3 + H_2O \qquad (1)$$

The crystal structure reveals that triphenylphosphane couples with both coordinated acetone molecules of **69** to form two molecules of the keto-stabilized ylide TPIP, part of the complex **70**. Complexes **69** and **70** are in equilibrium in acetone; addition of excess coordinating solvent to **70** results in the liberation of the TPIP and regeneration of **69**. No decomposition of the catalyst was observed even though the cycle reactions were performed numerous times. The excellent homogeneous catalysis may be ascribed to the presence of a heme-like metal Co^{II} centre, the electron-withdrawing effect of the fluorine groups (which enhances the Lewis acidity of the metal centre) and the effect of the tight fit of the TPIP inside the $F_{64}PcCo$ pocket of complex **70**.

5 Conclusion

We have reviewed recent progress on the synthesis, crystal structures and properties of monomeric Pcs. About 97 new Pc compounds with crystal structures have been synthesized in the past 5 years, based on several synthetic and modification strategies. Investigation of the crystal structures and supramolecular assemblies of these Pc compounds indicates that the introduction of axial and peripheral ligands can effectively hinder the Pc $\pi-\pi$ interactions, and even prohibit completely Pc $\pi-\pi$ overlapping. The extent of distortion and the shape of central Pc skeletons are dominated by the central metal ions, and by the number, steric effect and position of the substituents. As for ionic Pcs, it is noteworthy that the counterions usually form chain- or sheet-like structures located between the Pc $\pi-\pi$ stacks or ionic sheets. The UV–vis spectroscopic, conducting, magnetic and catalytic properties of monomeric Pcs have been discussed briefly. They are controlled mainly by the nature of the central metal ions, by the position, steric effects and electronic properties of the axis and the peripheral and axial substituted ligands, by the extent of distortion of Pc skeletons, by the arrangements of molecules in the solid state, and by the degree of oxidation of Pc rings, as well as by some other external measuring conditions.

Despite the vast number of reported syntheses, crystal structures, properties and applications of monomeric Pcs, it is still difficult to clarify clearly the synthetic mechanisms of Pcs in various conditions, to predict fully the supramolecular structures of Pcs in the solid state (except for a very few types of simple Pcs), and to elucidate completely the correlation between molecular structure and properties. The large-scale preparation and separation of some novel Pcs with interestingly properties is still very hard. On the basis of the great potential applications of Pcs in high-tech fields, exploitation of multi-functional Pc materials needs to be strengthened in the future. There is still plenty of room for further investigation of Pc chemistry.

Acknowledgments The authors thank the National Natural Science Foundation of China, the Education Ministry of China, and Shandong University for financial support.

References

1. Leznoff CC, Lever ABP (1996) Phthalocyanine: properties and applications, vols 1–4. Wiley-VCH, New York
2. McKeown NB (1998) Phthalocyanines materials: synthesis, structure and function. Cambridge University Press, New York
3. Yoshimoto S, Sawaguchi T, Su W, Jiang J, Kobayashi N (2007) Angew Chem Int Ed 46:1071
4. Torre G, Nicolau M, Torres T (2001) In: Nalwa HS (ed) Supramolecular photosensitive and electroactive materials. Academic, New York, p 1
5. Engel MK (2003) In: Kadish KM, Smith KM, Guilard R (eds) Porphyrin handbook. Academic, New York, p 1
6. Hanack M, Lang M (1994) Adv Mater 6:19
7. Riou MT, Clarisses C (1988) J Electroanal Chem 249:181
8. Schlettwein D, Wörhle D, Jaeger NI (1989) J Electrochem Soc 136:2882
9. Jiang J, Bao M, Rintoul L, Arnold DP (2006) Coord Chem Rev 250:424
10. Gregory P (1991) High-technology applications of organic colorants. Plenum, New York
11. Rawling T, McDonagh A (2007) Coord Chem Rev 251:1128
12. Inabe T (2001) J Porphyrins Phthalocyanines 5:3
13. Buchler JW, Ng DKP (2000) In: Kadish KM, Smith KM, Guilard R (eds) The porphyrin handbook, vol 3. Academic, San Diego, p 245
14. Jiang J, Kasuga K, Arnold DP (2001) In: Nalwa, HS (eds) Supramolecular photosensitive and electroactive materials. Academic, New York, p 113
15. Sosa-Sánchez JL, Sosa-Sánchez A, Farfn N, Zamudio-Rivera LS, López-Mendoza G, Flores JP, Beltrán HI (2005) Chem Eur J 11:4263
16. Beltrán HI, Esquivel R, Lozada-Cassou M, Dominguez-Aguilar MA, Sosa-Sánchez A, Sosa-Sánchez JL, Höfl H, Barba V, Luna-García R, Farfán N, Zamudio-Rivera LS (2005) Chem Eur J 11:2705
17. Liu W, Lee CH, Chan HS, Mak TCW, Ng DKP (2004) Eur J Inorg Chem 286
18. McKeown NB, Makhseed S, Msayib KJ, Ooi LL, Helliwell M, Warren JE (2005) Angew Chem Int Ed 44:7546
19. Bench BA, Sharman WM, Lee HJ, Gorun SM (2002) Angew Chem Int Ed 41:750
20. Matsuda M, Yamaura JI, Tajima H, Inabe T (2005) Chem Lett 34:1524
21. Wang R, Li R, Bian Y, Choi CF, Ng DKP, Dou J, Wang D, Zhu P, Ma C, Hartnell RD, Arnold DP, Jiang J (2005) Chem Eur J 11:7351
22. Bian Y, Wang R, Jiang J, Lee CH, Wang J, Ng DKP (2003) Chem Commun, p 1194
23. Sheng N, Li R, Choi CF, Su W, Ng DKP, Cui X, Yoshida K, Kobayashi N, Jiang J (2006) Inorg Chem 45:3794
24. Zhang H, Wang R, Zhu P, Lai Z, Han J, Choi CF, Ng DKP, Cui X, Ma C, Jiang J (2004) Inorg Chem 43:4740
25. Galezowski W, Kubicki M (2005) Inorg Chem 44:9902
26. Janczak J, Idemori YM (2002) Inorg Chem 41:5059
27. Janczak J, Kubiak R (2002) Polyhedron 21:265
28. Janczak J, Kubiak R, Śledź M, Borrmann H, Grin Y (2003) Polyhedron 22:2689
29. Janczak J, Kubiak R (2003) Inorg Chim Acta 342:64
30. Janczak J, Kubiak R (2007) Polyhedron 26:2997
31. Yang X, Kritikos M, Akermarka B, Sun L (2005) J Porphyrins Phthalocyanines 9:248
32. Cissell JA, Vaid TP, Rheingold AL (2005) Inorg Chem 45:2367
33. Fernández I, Pregosin PS, Albinati A, Rizzato S, Spichiger-Keller UE, Nezel T, Fernández-Sánchez JF (2006) Helv Chim Acta 89:1485
34. Brewis M, Helliwell M, McKeown NB (2003) Tetrahedron 59:3863
35. Farren C, FitzGerald S, Bryce MR, Beeby A, Batsanov AS (2002) J Chem Soc 59
36. Sosa-Sánchez JL, Galindo A, Gnecco D, Bernès S, Fern GR, Silver J, Sosa-Sánchez A, Enriquez RG (2002) J Porphyrins Phthalocyanines 6:198

37. Barker CA, Findlay KS, Bettington S, Batsanov AS, Perepichka IF, Bryce MR, Beeby A (2006) Tetrahedron 62:9433
38. Donzello MP, Bartolino L, Ercolani C, Rizzoli C (2006) Inorg Chem 45:6988
39. Kubiak R, Waśkowska A, Śledź M, Jezierski A (2006) Inorg Chim Acta 359:1344
40. Cui LY, Yang J, Fu Q, Zhao BZ, Tian L, Yu HL (2007) J Mol Struct 827:149
41. Janczak J, Idemori YM (2003) Polyhedron 22:1167
42. Wong A, Ida R, Mo X, Gan Z, Poh J, Wu G (2006) J Phys Chem A 110:10084
43. Kinzhybalo V, Janczak J (2007) Acta Cryst C 63:m357
44. Kinzhybalo V, Janczak J (2007) Inorg Chim Acta 360:3314
45. Guzei IA, McGaff RW, Kieler HM (2005) Acta Crys C 61:m472
46. Kubiak R, Waśkowska A, Pietraszko A, Bukowska E (2005) Inorg Chim Acta 358:453
47. Voloshin YZ, Varzatskii OA, Korobko SV, Chernii VY, Volkov SV, Tomachynski LA, Pehn'o VI, Yu M, Antipin ZA (2005) Inorg Chem 44:822
48. Beltrán HI, Esquivel R, Sosa-Sánchez A, Sosa-Sánchez JL, Höpfl H, Barba V, Farfán N, García MG, Olivares-Xometl O, Zamudio-Rivera LS (2004) Inorg Chem 43:3555
49. Liu W, Lee CH, Li HW, Lam CK, Wang J, Mak TCW, Ng DKP (2002) Chem Commun, p 628
50. Wang JD, Huang JL, Xu XZ, Cai JW, Chen NS (2004) Chin J Struct Chem 23:516
51. Bian Y, Li L, Dou J, Cheng DYY, Li R, Ma C, Ng DKP, Kobayashi N, Jiang J (2004) Inorg Chem 43:7539
52. Li R, Zhang Y, Zhou Y, Dong S, Zhang Y, Bian Y, Jiang J (2008) Cryst Growth Des (in press) doi: 10.1021/cg800342b
53. Wang JD, Huang JL, Cai JW, Chen NS (2001) Chin J Struct Chem 21:617
54. Donzello MP, Ercolani C, Gaberkorn AA, V Kudrik E, Meneghett M, Marcolongo G, Rizzoli C, Stuzhin PA (2003) Chem Eur J 9:4009
55. Lin MJ, Wang JD, Chen NS (2005) Z Anorg Allg Chem 631:1352
56. Gunaratne TC, Gusev AV, Peng X, Rosa A, Ricciardi G, Baerends EJ, Rizzoli C, Kenney ME, Rodgers MAJ (2005) J Phys Chem A 109:2078
57. Lin MJ, Wang JD, Chen NS, Huang JL (2006) Z Anorg Allg Chem 632:2315
58. Burnham PM, Cook MJ, Gerrard LA, Heeney MJ, Hughes DL (2003) Chem Commun, p 2064
59. Helliwell M, Teat SJ, Colesb SJ, Reeved W (2003) Acta Crystallogr B 59:617
60. Auger A, Burnham PM, Chambrier I, Cook MJ, Hughes DL (2005) J Mater Chem 15:168
61. Fukuda T, Homma S, Kobayashi N (2005) Chem Eur J 11:5205
62. McKeown NB, Helliwell M, Hassan BM, Hayhurst D, Li H, Thompson N, Teat SJ (2007) Chem Eur J 13:228
63. Enakieva YY, Gorbunova YG, Nefedov SE, Tsivadze AY (2004) Mendeleev Commun, p 193
64. Kimura T, Yomogita A, Matsutani T, Suzuki T, Tanaka I, Kawai Y, Takaguchi Y, Wakahara T, Akasaka T (2004) J Org Chem 69:4716
65. Lee HJ, Brennessel WW, Lessing JA, Brucker WW, Young VG, Gorun SM (2003) Chem Commun, p 1576
66. Bench BA, Beveridge A, Sharman WM, Diebold GJ, van Lier JE, Gorun SM (2002) Angew Chem Int Ed 41:748
67. Wang JD, Lin MJ, Wu SF, Lin Y (2006) J Organomet Chem 691:5074
68. Haas M, Liu SX, Neels A, Decurtins S (2006) Eur J Org Chem 5467
69. Gardberg AS, Yang S, Hoffman BM (2002) Inorg Chem 41:1778
70. Gardberg AS, Sprauve AE, Ibers JA (2002) Inorg Chim Acta 328:179
71. Gardberg AS, Deng K, Ellis DE, Ibers JA (2002) J Am Chem Soc 124:5476
72. Janczak J, Perpétuo GJ (2006) Acta Cryst C 62:m323
73. Matsuda M, Asari T, Naito T, Inabe T, Hanasaki N, Tajima H (2003) Bull Chem Soc Jpn 76:1935
74. Asari T, Naito T, Inabe T, Matsuda M, Tajimay H (2004) Chem Lett 33:128
75. Asari T, Ishikawa M, Naito T, Matsuda M, Tajima H, Inabe T (2005) Chem Lett 34:936
76. Yu DEC, Imai H, Ushio M, Takeda S, Naito T, Inabe T (2006) Chem Lett 35:602
77. Tutaß A, Klöpfer M, Hückstädt H, Cornelissen U, Homborg H (2002) Z Anorg Allg Chem 628:1027
78. Janczak J, Kubiak R (2001) Polyhedron 20:2901
79. Inabe T (2005) Bull Chem Soc Jpn 78:1373

Struct Bond (2009) 133: 161–206
DOI:10.1007/430_2008_15
© Springer-Verlag Berlin Heidelberg 2009
Published online: 5 March 2009

Controllable Assembly, Structures, and Properties of Lanthanide–Transition Metal–Amino Acid Clusters

Sheng-Chang Xiang, Sheng-Min Hu, Tian-Lu Sheng, Ling Chen, and Xin-Tao Wu

Abstract Amino acids are the basic building blocks in the chemistry of life. This chapter describes the controllable assembly, structures and properties of lathanide(III)–transition metal–amino acid clusters developed recently by our group. The effects on the assembly of several factors of influence, such as presence of a secondary ligand, lanthanides, crystallization conditions, the ratio of metal ions to amino acids, and transition metal ions have been expounded. The dynamic balance of metalloligands and the substitution of weak coordination bonds account for the occurrence of diverse structures in this series of compounds.

Keywords: 3d–4f cluster compounds · Amino acids · Properties · Self-assembly · Structures

Contents

1   Introduction 162
2   Coordination Chemistry and Binding Modes of Amino Acids 163
3   Effect of Secondary Ligand on the Assembly 166
   3.1   Heptanuclear Trigonal Prismatic Clusters Stabilized by Monodentate Imidazole Ligand 166
   3.2   Triacontanuclear Octahedral Clusters Stabilized by Bidentate Acetate Ligand 167
   3.3   Effect of Bidentate Acetate Ligand on the Assembly 169
4   Effect of Reactant Ratio on the Assembly 172
   4.1   1D $[La_6Cu_{24}(gly)_{14}(OH)_{30}(H_2O)_{24}(ClO_4)][Cu(gly)_2]_2 \cdot 21ClO_4 \cdot 26H_2O$ (**6**) 172
   4.2   2D $Na_2[Ln_6Cu_{24}(gly)_{14}(OH)_{30}(H_2O)_{22}(ClO_4)][Cu(gly)_2]_3 \cdot 23ClO_4 \cdot 28H_2O$ (**7**) (Ln = Eu, Gd and Er) 174

S.-C. Xiang, S.-M. Hu, T.-L. Sheng, L. Chen, and X.-T. Wu (✉)
State Key Laboratory of Structural Chemistry, Fujian Institute of Research on the Structure of Matter, Chinese Academy of Sciences, Fuzhou, Fujian 350002, P. R. China
e-mail: wxt@fjirsm.ac.cn

4.3	3D $[Sm_6Cu_{24}(OH)_{30}(gly)_{14}(ClO_4)(H_2O)_{22}]$ $[Cu(gly)_2]_5 \cdot 14ClO_4 \cdot 7OH \cdot$
	$24H_2O$ (**8**) 177
4.4	3D $[Nd_6Cu_{24}(OH)_{30}(pro)_{12}(ClO_4)(H_2O)_{21}]$ $[Cu(pro)_2]_6 \cdot 12ClO_4 \cdot 11OH \cdot$
	$6H_2O$ (**9**) 179

5   Effect of Crystallization Conditions on the Assembly 181
6   Effect of Lanthanide(III) Ions on the Assembly 185

6.1	$\{Ln(H_2O)_3[Cu(gly)_2][Cu(gly)_2(H_2O)]\}_2 \cdot 6ClO_4 \cdot 4H_2O$ (**11**)
	$(Ln^{3+} = La, Pr, Sm)$ 185
6.2	$Na_2[Eu_6Cu_{22}(OH)_{28}(Hgly)_4(gly)_{12}(ClO_4)(H_2O)_{18}][Cu(gly)_2]_3 \cdot 23ClO_4 \cdot$
	$28H_2O$ (**12**) and $[Dy_6Cu_{22}(OH)_{28}(Hgly)_4(gly)_{12}(ClO_4)(H_2O)_{18}][Cu(gly)_2]_3 \cdot$
	$21ClO_4 \cdot 20H_2O$ (**13**) 187
6.3	$Na_4[Er_6Cu_{24}(Hgly)_2(gly)_{12}(OH)_{30}(ClO_4)(H_2O)_{22}][Cu(gly)_2]_6 \cdot 27ClO_4 \cdot$
	$36H_2O$ (**14**) 190

7   Constitutionally Dynamic Chemistry of Cu–Ln–gly Compounds 193

| 7.1 | $Na_4[Tb_6Cu_{24}(OH)_{30}(gly)_{16}(ClO_4)(H_2O)_{18}][Cu(gly)(H_2O)_2]_2 \cdot 25ClO_4 \cdot$ |
| | $42H_2O$ (**15**) .... 196 |

8   Effect of Transition Metal on the Assembly 197

| 8.1 | $[LnM_6(AA)_{12}]^{3+}$ Octahedral Clusters ($M = Co^{2+}, Ni^{2+}$; $AA = $ gly, ala, thr) .... 198 |
| 8.2 | $[LnM_6]$ Trigonal Prismatic Clusters ($M = Co^{2+}, Ni^{2+}$, and Zn^{2+}) .... 199 |

9   Other Factors Affecting the Assembly 202
10   Conclusions 202
References 203

1 Introduction

The concept of self-assembly can be traced back to "coordination chemistry" pioneered by Werner. In 1987, Cram, Lehn, and Perdersen were awarded the Nobel Prize for laying the foundation of supramolecular chemistry. Through the self-assembly approach, many inorganic clusters with homo- [1–5] or heterometal [6–11], and complexes with organic ligand–metal coordination such as macrocycles [12], cages [13–15], polyhedra [16–18], and metal–organic frameworks (MOFs) [19–24] have been constructed. Herein, we present a summary of our recent results on the application of this approach to construct a wide range of heterometallic clusters with amino acids as ligands.

After several decades of intense development, the concept of "clusters" has been considerably expanded. Currently, the term cluster means a finite aggregate of atoms or molecules that are bound by forces that may be metallic, covalent, or ionic in character and can contain from a few to tens of thousands of atoms. Recently, sparked by the structural diversity and interesting properties found and proposed for 3d–4f clusters, interest in their rational syntheses has increased rapidly [25–28]. Firstly, 3d–4f heterometallic complexes have great potential as molecular magnets, especially single-molecule magnets [29–34], since 4f elements possess a large number of unpaired electrons and can introduce large magnetic anisotropy. Next, the introduction of a transition metal into a lanthanide complex may quench or increase the luminescence intensity of the lanthanide ion, which may have potential use in fluorescence devices and ion sensors [35–38]. Again, 3d–4f complexes can be used as precursors for the preparation of 3d–4f mixed-metal oxides, and many new superconducting materials can be synthesized [39, 40]. In order to synthesize a 3d–4f heterometallic compound, it is necessary to find a suitable ligand that can coordinate to lanthanide (Ln) and transition metal ions simultaneously. Previous works have proven that pyridones [41], Schiff bases [42], oxalato [43, 44], oximates [45, 46] or oxamides [47], cyano groups [48], and carboxylic acids [49] are good candidates for such purpose.

Although the self-assembly process is easy and convenient to operate, success in obtaining the expected object is still a challenge for chemists. The aims of this article are to summarize the coordination chemistry of amino acids, to review our recent work on 3d–4f heterometallic clusters bearing amino acid ligands, and to expound the effects of several factors of influence on self-assembly, such as presence of a secondary ligand, lanthanides, crystallization conditions, the ratio of Cu^{2+} to amino acids, and transition metal ions. We hope that our systematic researches on the 3d–4f amino acid clusters can provide a useful framework of reference for the study of other self-assembly systems.

2 Coordination Chemistry and Binding Modes of Amino Acids

Amino acid is one of the most important biological ligands. Researches on the coordination of metal–amino acid complexes will help us better understand the complicated behavior of the active site in a metal enzyme. Up to now many Ln–amino acid complexes [50] and 1:1 or 1:2 transition metal–amino acid complexes [51] with the structural motifs of mononuclear entity or chain have been synthesized. Recently, a series of polynuclear lanthanide clusters with amino acid as a ligand were reported (most of them display a Ln_4O_4-cubane structural motif) [52]. It is also well known that amino acids are useful ligands for the construction of polynuclear copper clusters [53–56]. Several studies on polynuclear transition metal clusters with amino acids as ligands, such as $[Co_3]$ [57, 58], $[Co_2Pt_2]$ [59], $[Zn_6]$ [60], and $[Fe_{12}]$ [61] were also reported.

Jacobson had obtained two homochiral Ni aspartate (asp) coordination polymers under hydrothermal conditions. One is a one-dimensional (1D) helical polymer containing a Ni–O–Ni infinite chain [63]. Each aspartate ligand is pentadentate, which is indicated by the Harris notation [63] as $[5.2_{12}2_{23}2_{34}2_{45}1_3]$. The other with a three-dimensional (3D) Ni–O–Ni connectivity forms at a higher pH and is based on the same helices as in the former chain polymer [64], which are connected by additional $[Ni(asp)_2]_2$-bridges to generate a chiral open framework containing 1D channels with minimum van der Waals dimensions of 8×5 Å. For both the 1D and 3D polymers, two transitions are observed at low temperature due to the infinite Ni–O–Ni connectivity. The 3D polymer undergoes a ferrimagnetic ordering transition with $T_c = 5.5$ K. Rosseinsky [65] also employed Ni aspartate units to generate homochiral porous materials with suitable bidentate linker molecules, which were used to probe enantioselective sorption for small chiral diol molecules. It was found that there is a geometry-dependent interaction between the nanoporous material and the small molecule guests.

For mixed lanthanide–transition metal clusters, Yukawa et al. have synthesized an octahedral $[SmNi_6]$ cluster by the reaction of Sm^{3+} and $[Ni(pro)_2]$ in nonaqueous medium [66–68]. The six $[Ni(pro)_2]$ "ligands" use 12 carboxylate oxygen atoms to coordinate to the Sm^{3+} ion, which is located at the center of an octahedral cage formed by six nickel atoms. The coordination polyhedron of the central Sm^{3+} ion may be best described as an icosahedron. The $[SmNi_6]$ core is stable in solution but the crystal is unstable in air. The cyclic voltammogram shows one reduction step from Sm^{3+} to Sm^{2+} and six oxidation steps due to the Ni^{2+} ions. Later, similar $[LaNi_6]$ and $[GdNi_6]$ clusters were also prepared.

A dinuclear Y–Cu complex was synthesized by the reaction of Y^{3+}, Cu^{2+}, and L-alanine in aqueous solution at relative low pH [69]. Four alaninato ligands use their carboxylate groups to bridge the two metal ions in the bidentate mode. The ligand acts more like a carboxylate than an amino acid because the amino group is protonated. Two 1D heterometallic coordination polymers $\{[CuEr(gly)_5(H_2O)_2] \cdot 5ClO_4 \cdot H_2O\}_n$ and $\{[Cu_2Gd_2(gly)_{10}(H_2O)_4] \cdot 10ClO_4 \cdot 4H_2O\}_n$ were also synthesized in a similar way [70]. Similar to those in heterometal carboxylates [71–73], the metal ions in these two compounds are all arranged in the linear mode with a $Ln \cdots Cu \cdots Cu \cdots Ln$ subunit. The carboxyl groups of glycine molecules take two coordination modes. The first one acts as a bidentate bridging ligand only. The second is a tridentate bridge that coordinates to three different metal ions. The amino groups of the glycinato ligands are also protonated.

Amino acids are ambidentate ligands with N-donor and O-donor atoms, which exhibit abundant coordination modes. As shown in Scheme 1, there are six main coordination modes (**a–f**) for the amino acids:

a. Amino acids can act as a "carboxylic acid ligand" to bridge two metal (transition metal or lanthanide) ions via two oxygen atoms from their carboxylic groups, which generally happens at low pH, in which case their amino groups are protonated [69].

b. When amino acids bind to transition metal ions, especially to divalent Cu, Ni, Zn, and Co ions with a metal-to-ligand ratio of 1:2, normally the amino group and one of the carboxylate oxygen atoms function in the chelate mode [51].

c. If there are remnant metal ions in the solution, they will further bind to another metal ion via the residual oxygen atom on the basis of mode **b** to form a chain [51] or ring compound (mode **c0**) [60]. Only when the remnant metals are lanthanide ions and free transition metal ions are absent (M:AA $\leq$ 1:2) can the residual oxygen atoms bind to Ln ions (mode **c1**). Mode **c2** is only seen in the case that a cis-Cu(AA)$_2$ group coordinates to one lanthanide ion via its two chelating carboxylate oxygen atoms.

d. If there are still a large amount of metal ions in solution, in the case of a high metal-to-ligand ratio, cluster compounds will be preferentially formed, and the oxygen atom taking part in the chelating interaction will further bind to a sodium or lanthanide ion on the basis of mode **c0** to form a cluster compound (modes **d0** and **d1**). Modes **d3** [62] and **d4** [52] have been observed only in compounds formed at high temperature and high pH conditions. In other words, it is difficult

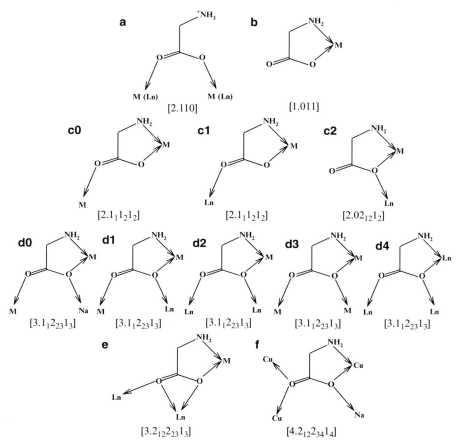

Scheme 1 Six main coordination modes (**a–f**) of the amino acid ligands indicated by the Harris notation, [63] where *M* is the transition metal ion

for an oxygen atom functioning in the chelating mode to further bind to another transition metal ion, and arduous for the amino group to chelate the lanthanide ion.

e. In the case of M:AA ≤ 1:2, the amino acid will further bind to another lanthanide ion on the basis of mode **c1** to form a cluster or chain compound via the chelating oxygen atom in the monodentate (mode **d2**) or chelating mode (mode **e**).

f. Occasionally, amino acids can further coordinate to another Cu ion via the nonchelating oxygen atom on the basis of mode **d0**, with a weak Cu–O bond length in the range 2.41–2.44 Å, to form a cluster dimer or condensed cluster [74]. Mode **f**, in which amino acids with three donors coordinated to a total of four metal centers is, to our knowledge, the largest.

From the discussions above, it can be concluded that amino acids prefer to chelate latter transition metals, then to bind to another metal ion via the nonchelating oxygen atom, while the oxygen atom already participating in chelation prefers binding to a Na^+ or Ln^{3+} ion.

In the following sections, our work on the amino acid clusters will be reviewed according to the effects of the various reaction conditions on self-assembly.

3 Effect of Secondary Ligand on the Assembly

In many cases, a secondary ligand is often employed to construct new compounds. For examples, 4,4'-bipyridine and its analogs are used to link two block units to form high-dimensional structures [75], while some clusters with higher nuclearities can be accessed by using secondary ligands, such as azido and halide anions, due to their bridging ligation and/or template effect [76–78]. Here, two secondary ligands, imidazole (im) and acetate, were employed to construct new 3d–4f clusters. They played a stabilization or interruption role in these novel cluster structures.

3.1 Heptanuclear Trigonal Prismatic Clusters Stabilized by Monodentate Imidazole Ligand

Three isomorphous heptanuclear trigonal prismatic complexes, $[LnCu_6(\mu_3-OH)_3(gly)_6im_6](ClO_4)_6$ (**1**) (Ln = La, Pr, Sm, Er) were synthesized through the self-assembly of $Ln(ClO_4)_3$, $Cu(ClO_4)_2$, glycine, and imidazole with a molar ratio of 1:6:6:6 in aqueous solution [79]. The complexes crystallize in the rhombohedral $R3$ space group. As shown in Fig. 1, six Cu^{2+} ions form a large trigonal prism with the

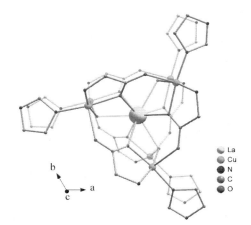

Fig. 1 Cationic structure of **1·La**

La^{3+} ion in the center. There are two parallel layers, each of which is composed of three Cu^{2+} ions. In each layer, the three Cu^{2+} ions form an equilateral triangle, and every pair of Cu^{2+} ions are connected by a chelating glycine ligand. Each Cu ion with a distorted pyramidal configuration has an N$_2$O$_3$ donor set that consists of one nitrogen atom of a glycine, one nitrogen atom of an imidazole, two carboxyl oxygen atoms from two glycinato ligands, and a μ_3–OH$^-$ group. The Ln^{3+} ion is surrounded by nine oxygen atoms to form a tricapped trigonal prismatic coordination polyhedron. Each glycinato ligand adopting a [3.1$_1$2$_2$3$_1$3] coordination mode (**d1** in Scheme 1) coordinates to two Cu^{2+} and one Ln^{3+} ion. Each imidazole is monodentate and coordinates to one Cu^{2+} ion. More interestingly, the ionic radii of the central lanthanide ions seem to have an effect on the distance of the two parallel layers and the angles of Cu–μ_3–OH$^-$–Cu. The distances for **1·La, 1·Pr, 1·Sm**, and **1·Er** are 3.497, 3.463, 3.432, and 3.385 Å, respectively, while the angles are 122.3, 121.2, 119.4, and 117.1° respectively. The trigonal prismatic clusters shrink in accordance with the contractive radii of the central lanthanide ions.

3.2 Triacontanuclear Octahedral Clusters Stabilized by Bidentate Acetate Ligand

With glycine or L-alanine as ligands, a series of novel 3d–4f heterometallic [Ln$_6$Cu$_{24}$] clusters, [Sm$_6$Cu$_{24}$(μ_3–OH)$_{30}$(gly)$_{12}$(Ac)$_{12}$(ClO$_4$)(H$_2$O)$_{16}$]·(ClO$_4$)$_9$· (OH)$_2$·(H$_2$O)$_{31}$ (**2**) and [Ln$_6$Cu$_{24}$(μ_3–OH)$_{30}$(ala)$_{12}$(Ac)$_6$(ClO$_4$)(H$_2$O)$_{12}$]·(ClO$_4$)$_{10}$· (OH)$_7$·(H$_2$O)$_{34}$ (**3**) (Ln = Tb, Gd, Sm, La) were synthesized in the presence of the acetate anion (Ac$^-$) [80]. The structure of **2** is shown in Fig. 2 as an example. The metal skeletons of the clusters are the same and may be described as a huge [Ln$_6$Cu$_{12}$] octahedron (the structure is almost the same as that reported by Chen et al. [81–85]) connected with 12 additional Cu^{2+} ions (every two are connected

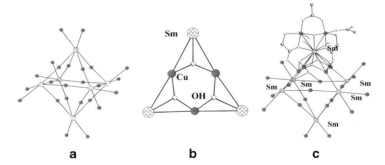

Fig. 2 Cationic structure of **2**. **a** Metal skeleton. **b** Structure of one of the faces of the octahedron. **c** View of the structure of one of vertices of the octahedron. Cu *filled circle*; Ln *circle with crosshatching*

to one Ln^{3+} vertex). Each Ln^{3+} ion interconnects two outer Cu^{2+} ions with the help of one outer μ$_3$–OH$^-$ and two [3.1$_1$2$_{23}$1$_3$]–coordinated glycinato ligands. The average Ln···Cu(outer) distance is about 3.5 Å, while that of Cu(outer)···Cu(outer) is 3.0 Å, shorter than that of Cu(inner)···Cu(inner). The coordination polyhedron of the nine-coordinated Sm^{3+} ion with an O$_9$ donor set may be best described as a monocapped square antiprism. A ClO$_4^-$ anion, which may play the template role, is encapsulated at the center of the octahedral cage. Previous authors have shown that a ClO$_4^-$ template is essential for the syntheses of [Ln$_6$Cu$_{12}$] clusters [81–85]. In contrast to [Ln$_6$Cu$_{12}$] carboxylates, each amino acid acting in the [3.1$_1$2$_{23}$1$_3$] mode employs its amino group and one carboxylate oxygen atom to chelate another outer Cu^{2+} ion, which increases the nuclearity of the cluster from 18 to 30. Another structural feature of the cluster is its size (with dimensions of about 2.38 × 2.38 × 2.38 nm^3), which is significantly larger than that of the famous [Mn$_{12}$] cluster [86] and comparable with that of the [Mn$_{30}$] cluster [87].

Temperature-dependent magnetic susceptibilities of complexes **3·La, 3·Nd, 3·Sm, 3·Gd, 3·Tb,** and **3·Dy** were measured as shown in Fig. 3 at an applied field of 5 kOe. At room temperature, the $\chi_M T$ values per Ln$_6$Cu$_{24}$ unit are 9.3, 72.8, and 56.2 cm^3 mol^{-1} K for **3·La, 3·Tb,** and **3·Gd**, respectively, as compared with the expected values (9.00, 79.88, and 56.25 cm^3 mol^{-1} K for **3·La, 3·Tb,** and **3·Gd**, respectively) for six Ln^{III} in the free-ion state and 24 spin-only Cu^{II} ions (S = 1/2, g = 2). Upon cooling, **3·La** shows a continuous decrease of $\chi_M T$, suggesting an overall antiferromagnetic coupling, as confirmed by the negative Weiss constant (−10.3 K). According to the literature [81], the Cu(inner)···Cu(inner) exchange interaction is antiferromagnetic. For the two neighboring outer Cu ions connected by a μ$_3$–OH$^-$ and a carboxylate group, as the ∠Cu(outer)–OH–Cu(outer) and the Cu(outer)···Cu(outer) distances are all about 100° and 3 Å, respectively, an antiferromagnetic interaction is also suggested [88]. **3·Sm** also shows an overall antiferromagnetic interaction. The free-ion approximation for Sm^{3+} is not valid because of the presence of thermally populated excited states. The $\chi_M T$ value at room temperature is about 11.2 cm^3 mol^{-1} K. Considering that in the isostructural **3·La** compound, the 24 Cu^{2+} ions have been shown to contribute 9.3 cm^3 mol^{-1} K to the

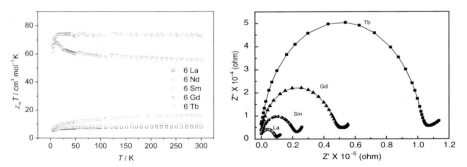

Fig. 3 For **3·Ln**, *left* temperature dependence of magnetic susceptibilities; *right* impendence spectroscopy of the electrical resistances at room temperature

bulk value, it can be deduced from the total $\chi_M T$ value that each Sm^{3+} ion accounts for $0.32\,cm^3\,mol^{-1}\,K$. This value is close to the expected value for an isolated non interacting Sm^{3+} ion [89]. The magnetic behavior of $3 \cdot Gd$ is different from those of $3 \cdot La$ and $3 \cdot Sm$. With decreasing temperature, $\chi_M T$ remains almost constant down to ca. 75 K, where it begins to increase smoothly until reaching a maximum of $68\,cm^3\,mol^{-1}\,K$ at around 5 K. This phenomenon corresponds to an overall ferromagnetic interaction and the Weiss constant determined in the range 50–300 K is $+1.9\,K$. This also indicates that the magnetic interaction of Gd–Cu is ferromagnetic. $3 \cdot Tb$ also shows an overall ferromagnetic interaction ($\theta = +0.59\,K$).

It is more interesting that these huge cluster compounds simultaneously exhibit magnetic and semiconducting properties. The impedance plot ($-Z''$ vs. Z') of $3 \cdot La$, $3 \cdot Sm$, $3 \cdot Gd$, and $3 \cdot Tb$ were recorded at room temperature ($22°C$) and are shown in Fig. 3. Using $3 \cdot Tb$ as an example, the measurement resulted in a typical behavior of ionic conductor with a semicircle at high frequencies (150–300 kHz) and a linear spike at low frequencies (20–150 Hz). From the plot, one can obtain the conductivities for $3 \cdot La$, $3 \cdot Sm$, $3 \cdot Gd$, and $3 \cdot Tb$ as 9.25×10^{-5}, 3.94×10^{-5}, 1.74×10^{-5}, and $7.72 \times 10^{-6}\,S\,cm^{-1}$, respectively. It is worth noting that only the organic charge transfer salts [90] with TTF or dmit units had been reported to exhibit magnetic molecular conductor properties before our work. Recently, Guo [91] employed transition-metal-complex-templated inorganic semiconductor to construct another new type of magnetic semiconductors.

3.3 Effect of Bidentate Acetate Ligand on the Assembly

With proline (pro) as the ligand, an interesting phenomenon was observed. The ratio of proline to acetate has a great effect on the 3d–4f prolinato clusters. The ratio of 1:4 will lead to triacontanuclear octahedral clusters similar to the triacontanuclear clusters above with alaninato or glycinato as ligand, while the ratio of 3:3 results in henhexacontanuclear clusters that comprises two triacontanuclear clusters connected by a $Cu(pro)_2$ linker. The large amount of acetate anions interrupt the interconnection between $Cu(pro)_2$ linkers and the triacontanuclear clusters and favor stabilization of the discrete triacontanuclear clusters.

In the presence of a number of acetate anions, a triacontanuclear mixed-metal prolinato cluster $[Tb_6Cu_{24}(OH)_{30}(pro)_{12}(ClO_4)(Ac)_9] \cdot 8ClO_4 \cdot 6(OH) \cdot 15.5H_2O$ (4) was obtained through the self-assembly process from a proline solution with a reactant ratio ($Ln^{3+}:Cu^{2+}:pro:Ac^-$) of 1:6:1:4. Single-crystal X-ray analysis of 4 reveals a highly symmetrical molecule that features crystallographically imposed $R\bar{3}m$ symmetry, in contrast to $P\bar{1}$ symmetry in compounds 2 and 3. As shown in Fig. 4, the metal skeletons of the cluster are similar to those of 2 and 3 and may be described as a huge $[Ln_6Cu_{12}]$ octahedron connected with 12 additional Cu^{II} ions (every two are connected to one Ln^{III} vertex), with a μ_{12}-ClO_4^- anion encapsulated at the center of the octahedral cage. Each Ln^{III} ion interconnects two outer Cu^{II} ions with the help of one outer μ_3-OH^- and two prolinato ligands. The $Ln \cdots Cu(outer)$ distances

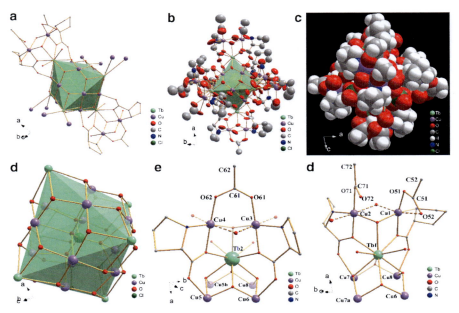

Fig. 4 Cationic structure of **4**. **a** Stereo view of the structure of the octahedron with only two of the vertices. **b** Ellipsoid and **c** space-filling representations of **4**. **d** Huge {Ln$_6$Cu$_{12}$} octahedral core Emcapsulates a ClO$_4^-$ anion. **e–f** Coordination environment of Tb^{3+} and the outer Cu^{2+} ions in the two vertices. The symmetry codes of *a* and *b* are z, x − 1, y+ 1; and y + 1, z − 1, x, respectively. The *dashed line* represents the weak coordination bonding

vary from 3.5358 to 3.5888 Å, while those of Cu(outer)···Cu(outer) are 3.0307 and 3.2336 Å, respectively. It is worth noting that in the known triacontanuclear heterometallic amino acids clusters (**2** and **3**) all vertices of the huge octahedron are stabilized by an acetate ion with the η$_2$ binding mode (Fig. 4e), while in compound **4** only half of the vertices are stabilized by an η$_2$-acetate ion and the other half by two acetate ions (Fig. 4f). The three acetate ligands adopt various binding modes. One is bidentate to coordinate two Cu^{2+} ions via its two O atoms, respectively; the second is monodentate to bind one Cu ion via one of its O atoms; the third is bidentate to bind one Cu ion via both of its O atoms. Each of the prolinato ligands in the [3.1$_1$2$_{23}$1$_3$] mode employs its amino group and one carboxylate oxygen atom to chelate another outer Cu^{2+} ion, which increases the nuclearity of the cluster from 18 to 30, in contrast to the [Ln$_6$Cu$_{12}$] carboxylates [81–85]. Without the consideration of weak coordination bonds, all outer Cu ions are each four-coordinated with an NO$_3$ donor set comprising a carboxylic O atom and an amino group from one prolinato ligand, a μ$_3$–OH group, and a carboxylic O atom from one acetate anion. The two outer Cu–μ$_3$–OH–Cu angles are 100.91 and 108.16°, respectively. All the inner Cu ions are six-coordinated by four μ$_3$–OH groups, one O atom from the encapsulated perchlorate anion, and one carboxylic O atom from the prolinato ligand. The distances of Cu(inner)···Cu(inner) range from 3.2951 to 3.4053 Å, and the corresponding Cu–μ$_3$–OH–Cu angles from 112.01 to 120.56°. The coordination polyhedron of the nine-coordinated Tb^{3+} ion with an O$_9$ donor set may be best described as a monocapped square antiprism. The dimensions of the cationic cluster are about 2.34 × 2.34 × 2.34 nm^3 (Fig. 4c).

For a Ln^{3+}:Cu^{2+}:pro:Ac$^-$ reactant molar ratio of 1:6:3:3, 61-nuclearity [Ln$_{12}$Cu$_{49}$] clusters {[Ln$_6$Cu$_{24.5}$(μ$_3$–OH)$_{30}$(pro)$_{13}$(Ac)$_6$ClO$_4$(H$_2$O)$_{13}$] · (ClO$_4$)$_9$ · (OH)$_8$ · (H$_2$O)$_{25.5}$}$_2$ (5) (Ln = Sm, Gd, Tb, Dy), rather than 30-nuclear [Ln$_6$Cu$_{24}$] clusters, can be obtained (Fig. 5) [92]. There are six crystallographically independent Ln^{3+} ions and 25 Cu^{2+} ions. The Cu13 atom residing at the 2a position is chelated by two prolinato ligands to form a *trans*-Cu(pro)$_2$ group (the L-prolinato ligand adopts the [2.1$_1$1$_2$1$_2$] coordination mode), and the other 24 Cu^{2+} ions and six Ln^{3+} ions are stabilized by 12 prolinato ligands with [3.1$_1$2$_{23}$1$_3$] mode and six acetate anions to form a [Ln$_6$Cu$_{24}$] cluster with a perchlorate template at its center. Compared with compound 4, each of the vertices of this huge [Ln$_6$Cu$_{24}$] cluster is stabilized by only one acetate anion with the η$_2$-binding mode (Fig. 5a, b). The [Ln$_{12}$Cu$_{49}$] cluster can be viewed as a dimer of the [Ln$_6$Cu$_{24}$] unit, and the structure can be described as two [Ln$_6$Cu$_{24}$] octahedral units connected by the *trans*-Cu(13)(proline)$_2$ group (Fig. 5c), which uses its carboxylate oxygen atoms

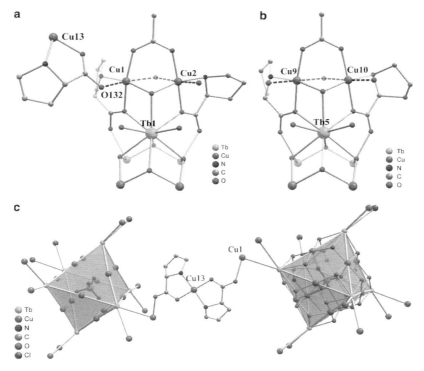

Fig. 5 a, b Coordination environment of Tb^{3+} and the outer Cu^{2+} ions at two different vertices. **c** Structure of 5·Tb

to bind to the six-coordinated sites of Cu1 ions. The $[Ln_{12}Cu_{49}]$ cluster also represents the largest known example of a 3d–4f heteronuclear cluster to date. An impressive structural feature of this cluster is its large size of $4.33 \times 2.38 \times 2.38\,nm^3$, which is significantly larger than that of other high-nuclearity 3d–4f heteronuclear clusters. The interconnection between $Cu(pro)_2$ linkers and the triacontanuclear $[Ln_6Cu_{24}]$ clusters is so weak (the distance Cu1-O132 is about 2.3 Å) that it can be interrupted by a large amount of acetate anions. Therefore, in order to improve the dimensionality of 3d–4f amino acid cluster compounds, the amount of ligands used, including acetate and the amino acids, should be carefully controlled because the η_2-coordination mode is also observed for the latter (mode **a** in Scheme 1).

4 Effect of Reactant Ratio on the Assembly

Generally, the reactant ratio will have an important effect on formation of the products. In the 3d–4f amino acid cluster system, we have also observed the effect of the reactant ratio on the assembly. In the case of a $Ln^{3+}:Cu^{2+}:gly$ ratio of 4:3:2, a 1D cluster compound (Ln = La) was obtained [93], while several 2D compounds (Ln = Eu, Gd and Er) were prepared in the case of 1:2:2 [94], and a 3D compound (Ln = Sm) in the case of 1:6:4 [95]. The most significant characteristic of these compounds is that the huge $[Ln_6Cu_{24}]$ clusters serve as nodes that are connected by $trans$-$Cu(gly)_2$ group linkers. Using the proline ligand, we also obtained a 3D compound with huge $[Nd_6Cu_{24}]$ clusters as the nodes and $trans$-$Cu(gly)_2$ groups as the linkers, with a reactant proportion $(Ln^{3+}:Cu^{2+}:pro)$ of 1:6:4 [95].

4.1 1D $[La_6Cu_{24}(gly)_{14}(OH)_{30}(H_2O)_{24}(ClO_4)]$ $[Cu(\,gly)_2]_2 \cdot 21ClO_4 \cdot 26H_2O$ (6)

There are three crystallographically independent La^{3+} ions, 13 Cu^{2+} ions, and nine glycinato ligands in $[La_6Cu_{24}(gly)_{14}(OH)_{30}(H_2O)_{24}(ClO_4)]$ $[Cu(gly)_2]_2 \cdot 21ClO_4 \cdot 26H_2O$ (**6**) [93]. The nine glycine ligands exhibit three coordination modes: (i) at the vertical orientation of the chain, one of the ligands chelates two Cu^{2+} ions through their carboxylate groups (**a** in Scheme 1); (ii) two of the ligands each bond to two Cu^{2+} ions through a nitrogen atom from the amino group and two oxygen atoms from their carboxylate group (**c0** in Scheme 1); (iii) the remaining ligands each chelate to two Cu^{2+} and one La^{3+} ions through their carboxylate and amino groups (**d1** in Scheme 1).

As shown in Fig. 6, each La^{3+} is coordinated by nine oxygen atoms, of which five are from μ_3–OH groups and the others from two water molecules and two glycine ligands. The 13 Cu^{2+} ions can be divided into three kind of coordination geometries: (i) six inner Cu^{II} centers each coordinated by six oxygen atoms in an octahedron; (ii) six outer and six-coordinated Cu^{II} centers showing an obvious Jahn–Teller

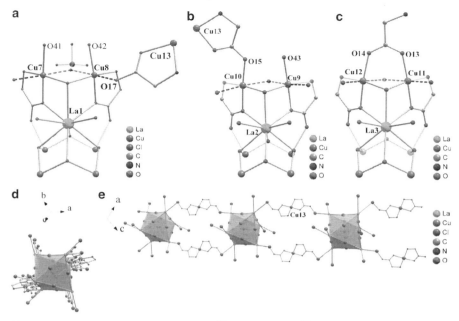

Fig. 6 a–c Coordination mode of the outer Cu^{2+} ions and the La^{3+} ions at the three vertices of the huge octahedral cluster. **d–e** Polymeric chain with {La$_6$Cu$_{24}$} clusters as nodes viewed along [111] and [010] directions, respectively

distortion effect; and (iii) one bridging Cu^{II} center in square–planar coordination involving two glycines through an amino nitrogen atom and a carboxylato oxygen to form the *trans*-Cu(gly)$_2$ group. Due to the distortion, the coordination environment of the outer Cu ions is an elongated octahedron in which the equatorial plane is completed by a μ$_3$–OH group, an amino group, a carboxylate oxygen atom from one glycinato ligand, and another oxygen donor. The last oxygen donor, which we designate as the "fourth coordinated" site in order to distinguish it from the first three coordinated sites, can be from the water molecule (the distances of Cu7–O41, Cu8–O42, and Cu9–O43 are 2.0147, 2.0085, and 1.9900 Å, respectively), the η$_2$-glycine ligands in the [2.110] mode (the distances of Cu11–O13 and Cu12–O14 are 1.9594 and 1.9661 Å, respectively), and even the *trans*-Cu(gly)$_2$ group (Cu10–O15 1.9422 Å). The respective dihedral angles of two quasi-planes of adjacent Cu(gly) fragments connected by μ$_3$–OH group are about 108.9° in the case of adjacent Cu^{2+} ions bridged by the η$_2$-glycine ligand (Fig. 6c), and about 116.0° in cases without the η$_2$-bridge (Fig. 6a, b). The "fifth coordinated" sites of the outer Cu^{2+} ions simultaneously connecting two adjacent quasi-planes are from the water and perchlorate ligands, with Cu–O distances ranging from 2.427 to 2.612 Å. The residual "sixth coordinated" sites of the outer Cu^{2+} ions can be an oxygen atom from the water molecule with Cu–O distances in the range of 2.35–2.59 Å, or from the *trans*-Cu(gly)$_2$ group (Cu8–O17 2.2574 Å). The three La^{3+} ions, 12 inner and outer Cu^{2+}

ions, and seven glycinato ligands generate a centrosymmetric 30-nuclearity cluster, with a perchlorate anion located at the crystallographic inversion center. In the cluster, the inner core has pseudo-O_h symmetry with six La^{3+} ions positioned at the vertices of a regular octahedron and 12 Cu^{2+} ions located at the midpoints of the edges. Each La^{3-} ion is connected to four Cu^{2+} ions by four μ_3–OH groups in the octahedron and to two Cu^{2+} ions by one exohedral μ_3–OH. In the octahedron, each μ_3–OH links one La^{3+} ion and two Cu^{2+} ions. The average La–Cu and Cu–Cu distances are 3.579 and 3.398 Å, respectively, and the average La–Cu–La angle is 174.0°. Each La^{3+} is also connected to two exotedral Cu^{2+} by two glycines with the $[3.1_122_313]$ binding mode, and a Cu–La–Cu angle of 97.5°. The *trans*-Cu(13)(gly)$_2$ group employs its two oxygen atoms O15 and O17 to respectively occupy the fourth coordinated site of the Cu10 ion and the six-coordinated site of the Cu8 ion of the adjacent cluster in the next cell. As a result, a double-chain polymeric structure along the [111] direction is formed, as shown in Fig. 6d, e. Other than compound **6**, Chen [81–85] and Winpenny [96] have respectively reported the high-nuclearity 3d–4f clusters $[Cu_{12}Ln_6(\mu_3–OH)_{24}(ClO_4)]^{17+}$ and $[Cu_{12}Ln_8(\mu_3–OH)_{24}(NO_3)]^{23+}$. The other known example with an amino acid ligand is $[Ln_{15}Tyr_{10}(OH)_{32}Cl]^{2+}$, reported by Zheng [52].

The magnetic susceptibility data for compound **6** were collected from a polycrystalline sample at an external field of 10 kOe on a Quantum Design PPMS Model 6000 magnetometer in the temperature range of 5–300 K. The data were corrected for experimentally determined diamagnetism of the sample holder, and the diamagnetism of the samples was calculated from Pascal's constants. $\chi_M T$ is 9.24 cm^3 K mol^{-1} at 300 K, which slightly increases from 300 to 100 K. As T is further lowered, the $\chi_M T$ value increases and reaches 14.77 cm^3 K mol^{-1} at 5.0 K. The $\chi_M T$ value at room temperature (T at 300 K) is comparable to the calculated value using Eq. 1, which is based on the free-ion approximation for Cu^{II}, where N_A, β and k are the Avogadro number, the Bohr magneton, and the Boltzmann constant, respectively:

$$\chi_M T = [N_A\beta^2\{\Sigma g^2_{Cu(II)}(S_{Cu(II)}(S_{Cu(II)} + 1))\}]/3k \tag{1}$$

The magnetic data of the polymeric complex at low temperature ($T < 20$ K) yield much higher values than those calculated using Eq. 1 and show strong ferromagnetic interactions. The magnetic behavior of the complex is different from those of discrete Cu–Ln complexes, which normally show antiferromagnetic interactions [81–85].

4.2 2D Na$_2$[Ln$_6$Cu$_{24}$(gly)$_{14}$(OH)$_{30}$(H$_2$O)$_{22}$(ClO$_4$)][Cu(gly)$_2$]$_3$ · 23ClO$_4$ · 28H$_2$O (7) (Ln = Eu, Gd and Er)

The three compounds are isostructural, and only the structure of **7** · **Er** is described here. The most notable feature of **7** · **Er** is a 2D network composed of Eu$_6$Cu$_{24}$ cluster nodes and *trans*-Cu(gly)$_2$ group linkers, as shown in Fig. 7e.

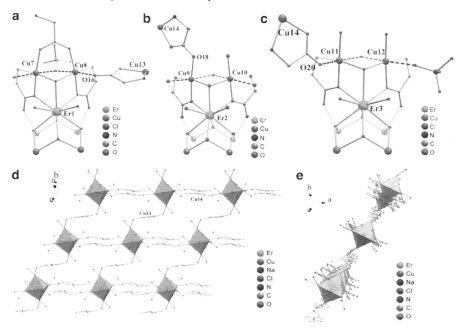

Fig. 7 a–c Coordination mode of the outer Cu^{2+} ions and the Er^{3+} ions at the three vertices of the huge octahedral cluster for **7·Er**. **d–e** 2D layer of **7·Er** with $\{Er_6Cu_{24}\}$ clusters as nodes viewed along [1–10] and [111] directions, respectively

There are three crystallographically independent Er^{3+} ions, 14 Cu^{2+} ions and ten glycinato ligands in $Na_2[Er_6Cu_{24}(gly)_{14}(OH)_{30}(H_2O)_{22}(ClO_4)][Cu(gly)_2]_3 \cdot 23ClO_4 \cdot 28H_2O$ (**7·Er**). Similar to compound **6**, each Er^{3+} here is coordinated by nine oxygen atoms, of which five are from μ_3–OH groups and the others from two water molecules and two glycine ligands. The 14 Cu^{2+} ions can be divided into three kinds of coordination geometries: (i) six inner Cu^{2+} centers each coordinated by six oxygen atoms in an octahedron; (ii) six outer Cu^{2+} centers exhibiting an obvious Jahn–Teller distortion effect are coordinated by a NO_5 donor set; and (iii) two bridging Cu^{2+} centers in square–planar coordination mode each linked by two glycines through an amino nitrogen atom and a carboxylato oxygen to form the *trans*-$Cu(gly)_2$ group. The $[Er_6Cu_{24}]$ octahedral-like node may be described as a huge $[Er_6Cu_{12}]$ octahedron (inner core) with pseudocubic O_h symmetry and 12 outer Cu^{2+} ions. Six Er^{3+} ions are located at the vertices and 12 inner Cu^{2+} ions occupy the midpoints of each edge of the octahedron with an edge distance of about 6.92 Å. The average Er···Cu(inner) and Cu(inner)···Cu(inner) distances are about 3.46 and 3.36 Å, respectively. Each Er^{3+} ion interconnects two outer Cu^{2+} ions with the help of one outer μ_3–OH$^-$ and two [3.1$_1$2$_{23}$1$_3$]-coordinated glycine ligands. The average Er···Cu(outer) distance is about 3.53 Å, while that of two neighboring outer Cu^{2+} ions is about 3.14 Å, being shorter than the Cu(inner)···Cu(inner) contact. The 30 μ_3–OH$^-$ groups, each linking one Er^{3+}

and two Cu^{2+} ions, were used to construct the $[Er_6Cu_{24}]$ cluster node whose central cavity encapsulates a μ_{12}–ClO_4^- as the template.

Figure 7 shows the 2D net-like structure of **7·Er**. There are two types of *trans*-Cu(gly)$_2$ groups: (i) *trans*-Cu(14)(gly)$_2$, which employs its two oxygen atoms O18 and O20 to respectively occupy the fourth coordinated site of Cu9 ion and the six-coordinated site of Cu11 ion of the adjacent cluster in the next cell; (ii) Cu13 positions at $1f$ site and its *trans*-Cu(gly)$_2$ groups employ two crystalographically dependent O18 atoms to occupy the six-coordinated sites of two Cu11 ions of adjacent clusters in the next cell. In the crystal, each $[Er_6Cu_{24}]$ unit is firstly polymerized through two *trans*-Cu(14)(gly)$_2$ bridges to yield a double-chain running parallel to the [111] direction. Then these chains are further connected by one *trans*-Cu(13)(gly)$_2$ group. That is, each $[Er_6Cu_{24}]$ unit is connected through six *trans*-Cu(gly)$_2$ bridges to four neighboring $[Er_6Cu_{24}]$ units, resulting in a 2D net-like structure extended along the [111] and [001] directions.

In fact, the structure of the $[Ln_6Cu_{12}]$ inner core is similar to the $[Ln_6Cu_{12}]$ cluster containing μ_2-coordinated betaine. But, as the amino acid has more coordination modes than betaine, this makes the structure of our complex much more intriguing than the discrete 18-nuclearity complex:

1. [2.110]-coordinated glycines, each coordinates to neighboring outer Cu^{2+} ions.
2. [3.1$_1$2$_{23}$1$_3$]-coordinated glycines, each chelates to one inner Cu^{2+}, one outer Cu^{2+}, and one Er^{3+} ions. Twelve more Cu^{2+} ions are introduced into the system, and thus a higher-nuclearity cluster is obtained.
3. [2.1$_1$1$_2$1$_2$]-coordinated glycines, two of which coordinate to one bridge Cu^{2+} ion. The *trans*-Cu(gly)$_2$ linker thus formed is used to bridge high-nuclearity nodes to yield the 2D polymer.

The length of the *trans*-Cu(gly)$_2$ linker (the distance of the two spare carboxylate oxygen atoms) is about 7.83 Å, as compared with 7.34 Å of terephthalic acid and 7.08 Å of 4,4′-bipyridine. This linker uses two spare carboxylate oxygen atoms to coordinate to the outer Cu^{II} of the $[Er_6Cu_{24}]$ unit, and the two nodes are thus bridged.

The electrical conductivity of **7·Er** was determined using a powder sample from ground crystals. The electrical conductivity of **7·Er** is about $1.25 \times 10^{-7}\,S\,cm^{-1}$ at 238.15 K and increases as the temperature rises, which indicates that it is a semiconductor. The temperature-dependent magnetic susceptibility of complex **7·Er** was also measured. At room temperature, the measured $\chi_M T$ value is $80.97\,cm^3\,mol^{-1}\,K$, as compared with the expected value of 78.98. Upon cooling, the complex shows a continuous decrease in $\chi_M T$, suggesting an overall antiferromagnetic coupling, as confirmed by the negative Weiss constant (-6.9 K). According to the literature, the Cu(inner)$\cdots$Cu(inner) and Cu(outer)$\cdots$Cu(outer) exchange interactions are all antiferromagnetic, but the Cu (bridge)$\cdots$Cu (outer) exchange interaction may be weakly ferromagnetic. The overall antiferromagnetic interaction of the complex also indicates that the magnetic interaction of Er–Cu may be antiferromagnetic.

4.3 3D [Sm₆Cu₂₄(OH)₃₀(gly)₁₄(ClO₄)(H₂O)₂₂][Cu(gly)₂]₅ · 14ClO₄ · 7OH · 24H₂O (8)

The complex $[Sm_6Cu_{24}(OH)_{30}(gly)_{14}(ClO_4)(H_2O)_{22}][Cu(gly)_2]_5 \cdot 14ClO_4 \cdot 7OH \cdot 24H_2O$ (**8**) possesses a 3D network based on $[Sm_6Cu_{24}]$ nodes and *trans*-$Cu(gly)_2$ bridges. There are three crystallographically independent Sm^{3+} ions, 15 Cu^{2+} ions, and 12 glycinato ligands in compound **8**. Each Sm^{3+} here is coordinated by nine oxygen atoms, of which five are from μ_3–OH groups and the others from two water molecules and two glycine ligands. The 15 Cu^{2+} ions can be divided into three kind of coordination geometries: (i) six inner Cu^{II} centers each coordinated by six oxygen atoms in an octahedron; (ii) six outer Cu^{II} centers exhibiting obvious Jahn–Teller distortion are coordinated by a NO_5 donor set; and (iii) three bridging Cu^{II} centers in square–planar coordination mode, which are each linked by two glycines through an amino nitrogen atom and a carboxylato oxygen to form the *trans*-$Cu(gly)_2$ group. The $[Sm_6Cu_{24}]$ octahedral-like node may be described as a huge $[Sm_6Cu_{12}]$ octahedron (inner core) with pseudocubic O_h symmetry and 12 outer Cu^{II} ions. Six Sm^{3+} ions are located at the vertices and 12 inner Cu^{II} ions are located at the midpoints of each edge of the octahedron with an edge distance of about 7.00 Å. The average $Sm \cdots Cu(inner)$ and $Cu(inner) \cdots Cu(inner)$ distances are about 3.46 and 3.37 Å, respectively. Each Sm^{3+} ion interconnects two outer Cu^{2+} ions with the help of one outer μ_3–OH^- and two $[3.1_12_{23}1_3]$-coordinated glycine ligands. The average $Sm \cdots Cu(outer)$ distance is about 3.57 Å, while that of two neighboring outer Cu^{2+} ions is about 3.19 Å, being shorter than that of $Cu(inner) \cdots Cu(inner)$. Thirty μ_3–OH^- groups, each linking one Sm^{3+} and two Cu^{2+} ions, are used to construct the $[Sm_6Cu_{24}]$ cluster node at the center of which a μ_{12}–ClO_4^- anion is encapsulated as the template.

Figure 8a–c shows the structures of the three vertices of the octahedron. Among the six outer Cu^{2+} ions, only Cu11 and Cu12 ions are bridged by an η_2-coordinated glycine, while the fourth coordinated sites of Cu8 and Cu10 ions are occupied by water molecules, and those of Cu7 and Cu9 ions by oxygen atoms from the *trans*-$Cu(gly)_2$ groups (the bond lengths of Cu7–O14 and Cu9–O19 are 1.9841 and 1.9583 Å, respectively). The sixth coordinated sites of Cu8, Cu11, and Cu12 ions are occupied by the oxygen atoms from the *trans*-$Cu(gly)_2$ groups with corresponding Cu–O distances of 2.3570, 2.3262, and 2.3678 Å, respectively. The other sites of the six outer Cu^{2+} ions are similar to those in the compound **6**. Among the three Cu ions that form the *trans*-$Cu(gly)_2$ groups, Cu15 ion sites at the $1a$ position and its *trans*-$Cu(gly)_2$ groups link adjacent $[Sm_6Cu_{24}]$ clusters to form a chain along the [10-1] direction. The other Cu ions (Cu13 and Cu14) connect the adjacent $[Sm_6Cu_{24}]$ cluster nodes to form double-chains along the [001] and [010] directions, respectively (Fig. 8d). The two double-chains further crossover to form a plane extending along the *bc* plane. The *trans*-$Cu(15)(gly)_2$ groups link the planes to form a 3D open framework (Fig. 8e). Within the framework, each $[Sm_6Cu_{24}]$ unit is connected through ten *trans*-$Cu(gly)_2$ bridges to six neighboring $[Sm_6Cu_{24}]$ units, and the network topology might be described as distorted primitive cubic (Fig. 8f),

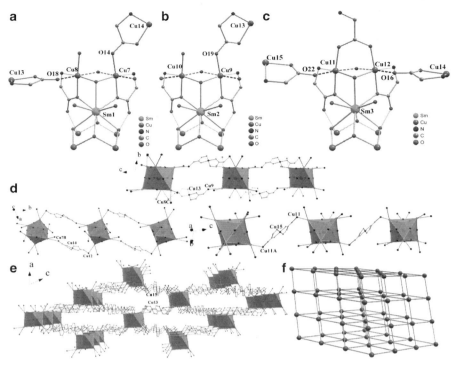

Fig. 8 a–c Coordination mode of the outer Cu^{2+} ions and the Sm^{3+} ions at the three vertices of the huge octahedral cluster for **8**. **d** $[Sm_6Cu_{24}]$ cluster nodes linked by the three *trans*-$Cu(gly)_2$ groups to generate the chains along [001], [010], and [10-1] directions, respectively. Symmetry codes for A, B, and C are $2-x, 2-y, z$; $x, y-1, z$; and $x, y, 1+z$, respectively. **e** 3D open-framework of **8** viewed along **b** axis. **f** Primitive cubic network

with the "brickwall"-like structure and rectangle channels running parallel to the b directions. The rectangular channel has dimensions of about $7 \times 31 \text{Å}^2$. The effective free volume of **8** is about 3976Å^3, comprising 56.8% of the crystal volume, as calculated by the program PLATON. This value is large among known microporous networks and close to that observed in the 3D supramolecule with $[Cd_8]$ as nodes [97]. Free water molecules, perchlorate ions and hydroxides are encapsulated in the large pores.

As a number of free water molecules, perchlorate ions, and hydroxides fill its channels, compound **8** exhibits the behavior of an ionic conductor. The electrical conductivities of **8** were determined using a powder sample from ground crystals. The electrical conductivity of **8** at 263.15 K is 1.72×10^{-4} S cm^{-1} and increases to 2.57×10^{-3} S cm^{-1} at 318.15 K, which indicates that **8** is a semiconductor. In addition, temperature-dependent magnetic susceptibilities of complexes **8** were measured in the range 2–300 K at 2000 and 5000 G, respectively. Antiferromagnetic interactions were observed for **8**, as confirmed by the Weiss constants of -43.7 K. According to the literature [81–85], the Cu(inner)···Cu(inner) exchange interaction is antiferromagnetic. The two neighboring outer Cu ions are connected by a μ_3–OH$^-$ and a carboxylate group, and as the $\angle$Cu(outer)–OH–Cu(outer) angle and the Cu(outer)$\cdots$Cu(outer) distance are about 100° and 3 Å, respectively, an antiferromagnetic interaction is also suggested [88]. The Cu (bridging)$\cdots$Cu (outer) distance is about 5.3 Å, and a glycine ligand is used to connect them in the η_2-coordination mode (*syn–anti*). According to the literature [98], a weak ferromagnetic coupling occurs when two Cu ions are connected by a carboxylate group in the *syn–anti* coordination mode. Thus, a similar weak ferromagnetic interaction is also suggested between the bridging and outer Cu ions in **8**.

4.4 3D [Nd$_6$Cu$_{24}$(OH)$_{30}$(pro)$_{12}$(ClO$_4$)(H$_2$O)$_{21}$][Cu(pro)$_2$]$_6$ · 12ClO$_4$ · 11OH · 6H$_2$O(9)

When the chiral amino acid proline was used as the ligand instead of glycine, a 3D [Nd$_6$Cu$_{24}$(OH)$_{30}$(pro)$_{12}$(ClO$_4$)(H$_2$O)$_{21}$][Cu(pro)$_2$]$_6$ · 12ClO$_4$ · 11OH · 6H$_2$O(**9**), which crystallized in the chiral $P2(1)3$ space group, could be obtained under the same reaction conditions as compound **8**.

Complex **9** is a 3D network based on [Nd$_6$Cu$_{24}$] nodes and *trans*-Cu(pro)$_2$ bridges and represents a rare example of constructing a chiral framework from simple reagents and reaction. There are two crystallographically independent Nd^{3+} ions, ten Cu^{2+} ions and eight prolinato ligands in compound **9**. Each Nd^{3+} here is coordinated by nine oxygen atoms, of which five are from μ_3–OH groups and the others from two water molecules and two glycine ligands. The ten Cu^{2+} ions can be divided into three kinds of coordination geometries: (i) four inner Cu^{2+} centers that are each coordinated by six oxygen atoms in an octahedron; (ii) four outer Cu^{2+} centers with an obvious Jahn–Teller distortion effect; (iii) and two bridging Cu^{2+} centers in square–planar coordination mode, which are each linked by two prolines through an amino nitrogen atom and a carboxylato oxygen to form the *trans*-Cu(pro)$_2$ group. The [Nd$_6$Cu$_{24}$] octahedral-like node may be described as a huge [Nd$_6$Cu$_{12}$] octahedron (inner core) with pseudocubic O_h symmetry and 12 outer Cu^{2+} ions. Six Nd^{3+} ions are located at the vertices and 12 inner Cu^{2+} ions are located at the midpoints of each edge of the octahedron with an edge distance of about 7.08 Å. The average Nd$\cdots$Cu(inner) and Cu(inner)$\cdots$ Cu(inner) distances are about 3.54 and 3.44 Å, respectively. Each Nd^{3+} ion interconnects two outer Cu^{2+} ions with the help of one outer μ_3–OH$^-$ and two [3.1$_1$2$_{23}$1$_3$]-coordinated prolinato ligands. The average Nd$\cdots$Cu(outer) distance is about 3.61 Å, while that of two neighboring outer Cu^{2+} is about 3.25 Å, shorter than that of Cu(inner)$\cdots$Cu(inner). The 30 μ_3–OH$^-$ groups, each one linking one Nd^{3+} and two Cu^{2+} ions, are used to construct the [Nd$_6$Cu$_{24}$] cluster node, at the center of which a μ_{12}–ClO$_4^-$ anion is encapsulated as the template.

Figure 9a, b shows the structures of two vertices of the octahedron. Among the four crystallographically independent outer Cu^{2+} ions, the Cu5 ion is five-coordinated in square–pyramidal geometry and has an NO$_4$ donor set that con-

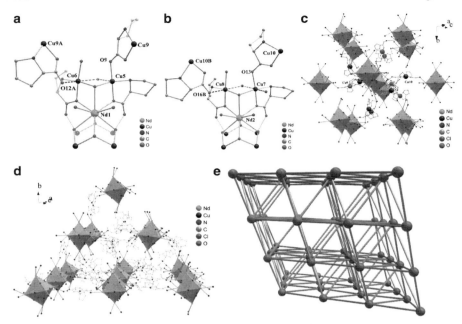

Fig. 9 a–b Coordination mode of the outer Cu^{2+} ions and the Nd^{3+} ions at the two vertices of the huge octahedral cluster {Nd$_6$Cu$_{24}$} for **9**. Symmetry codes for A and B are y, z, x; and 0.5 − z, 1 − x, −0.5 + y, respectively. **c** Each cluster nodes link to 12 other cluster units through 12 *trans*-Cu(pro)$_2$ groups. **d** 3D open-framework of **9**. **e** Face-centered cubic network

sists of one amino nitrogen atom and one carboxylate atom from proline, one outer μ$_3$–OH$^−$, one carboxylate oxygen from one *trans*-Cu(pro)$_2$ bridge, and one water molecule. The other outer Cu^{2+} ions (Cu6, Cu7, and Cu8) are six-coordinated in square–pyramidal geometry and have an NO$_5$ donor set that consists of one amino nitrogen atom and one carboxylate atom from proline, one outer μ$_3$–OH$^−$, one carboxylate oxygen from one *trans*-Cu(pro)$_2$ bridge, and two water molecules. The difference between them is that the fourth coordinated site of Cu7 ion, similar to that of Cu5 ion (Cu5–O9 1.9512 Å), is occupied by the oxygen atom from the *trans*-Cu(pro)$_2$ group (Cu7–O13 1.9887 Å), while the oxygen atoms from the *trans*-Cu(pro)$_2$ group occupy the sixth coordinated sites of Cu6 and Cu8 ions (Cu6–O12A 2.3753 Å; Cu8–O16B 2.3998 Å). Compared with the glycinato ligands in compounds **6–8**, the prolinato ligand with the η$_2$-coordination mode that was observed in the Fe$_{12}$ cluster [61] does not occur in compound **9**. There are only two types of prolinato ligands in compound **9**: (i) [3.1$_1$2$_{23}$1$_3$]-coordinated prolines, each of which chelates to one inner Cu^{II}, one outer Cu^{2+} and one Nd^{3+} ion so that a higher-nuclearity cluster node is obtained; (ii) [2.1$_1$1$_2$1$_2$]-coordinated prolines, two of which coordinate to one bridging Cu^{2+} ion to form a *trans*-Cu(pro)$_2$ group, which is used to link high-nuclearity cluster nodes to obtain the 3D polymer.

The steric effect of the L-proline side chain, compared with glycine, is responsible for the large structural difference between compound **8** and **9**. In **9**, each $[Nd_6Cu_{24}]$ node is connected to 12 neighboring $[Nd_6Cu_{24}]$ units with the help of 12 *trans*-Cu(pro)$_2$ linkers; (Fig. 9c) the structure might be described as a cubic close-packed network (also known as face-centered cubic), a type of packing of prime importance in crystallography (as shown in Fig. 9e). From another point of view, compound **9** may also be viewed as constructed from $[Nd_6Cu_{24}]$ tetrahedral building blocks with an edge of about 23 Å (the distance between the ClO_4^- atoms captured in the metal cage), as shown in Fig. 9d. This block not only has a large pore itself, but also can form superlattices with large pore size and high pore volume compared with the close-packed lattices of the colloidal nanoparticles. Complex **9** represents a very rare example of transition metal coordination polymer constructed from high-nuclearity tetrahedral building blocks along with the chalcogenide supertetrahedral frameworks [99].

The effective free volume of **9** is about $16,283\,\text{Å}^3$, comprising 47.7% of the crystal volume, as calculated by the program PLATON. Free water molecules, hydroxide ions, and ClO_4^- ions are encapsulated in the large pores, which leads to the semiconductivity of compound **9**. The electrical conductivity of **9** is $4.27 \times 10^{-7}\,\text{S cm}^{-1}$ at 273.15 K and increases to about $6.84 \times 10^{-6}\,\text{S cm}^{-1}$ at 310 K, as determined using a powder sample from grounded crystals. The difference in the electrical conductivity between **8** and **9** indicates that the packing mode of The $[Ln_6Cu_{24}]$ building block might have a great influence. Temperature-dependent magnetic susceptibilities of complexes **9** were measured in the range 2–300 K at 2000 and 5000 G. Similar to that in compound **8**, antiferromagnetic interaction was also observed for **9**, as confirmed by the Weiss constants of -38.2 K.

5 Effect of Crystallization Conditions on the Assembly

All the crystalline compounds reported above were separated from the mother liquor in a desiccator filled with phosphorus pentaoxide. Recently, a new phenomenon was observed that colorless solid substances, rather than the blue crystals, were first precipitated from a parent solution containing reactants $(Gd^{3+} : Cu^{2+} : \text{gly})$ of molar ratio of 2:1:2 at a pH value of about 6.6, which was placed in a desiccator filled with phosphorus pentoxide. The colorless solid is very bibulous and noncrystalline, which is probably a Gd–oxo perchlorate. Then the flask containing the mixtures was moved out of the desiccator and quietly placed in air. Blue crystals of $[Gd(H_2O)_8] \cdot [Gd_6Cu_{12}(OH)_{14}(gly)_{15}(Hgly)_3(H_2O)_6] \cdot 16ClO_4 \cdot 14H_2O$ (**10**) [100] were formed and isolated after 2 weeks. Experimental evidence based on the elemental analysis and energy dispersive X-ray spectra confirmed the molecular formula of the compound. The powder X-ray diffraction measurement shows the pure phase of the samples. If the initial step of precipitation of the colorless solid was omitted, compound **10** could not be separated. The present reaction suggests that a stepwise growth process takes place, whereby the initial colorless solid acts as a matrix for the aggregation of additional building units to form compound **10**.

Single-crystal X-ray analysis of **10** reveals a highly symmetrical molecule that features crystallographically imposed $\bar{3}$ symmetry. Compound **10** is composed of a discrete octadecanuclear cation $[Gd_6Cu_{12}(OH)_{14}(gly)_{15}(Hgly)_3(H_2O)_6]^{13+} \equiv$ $[Gd_6Cu_{12}]$, an octa-aqua Gd^{3+} ion, several perchlorates, and lattice water molecules (Fig. 10). There are one crystallographically independent Gd and two Cu atoms, as well as three gly ligands, in the octadecanuclear cation (Fig. 10a). The Cu1 and Cu2 ions adopt four-coordinated NO_3 square–planar and five-coordinated NO_4 square–pyramidal geometry, respectively, while the Gd1 ion adopts a nine-coordinated O_9 monocapped square antiprismatic geometry. The two Cu atoms are linked to form a binuclear copper-gly fragment $[Cu_2(gly)_3(OH)(H_2O)] \equiv [Cu_2]$ by one μ_3–O(3)H ligand as well as one gly in a *syn–syn* binding mode. Six symmetry-related Gd1 atoms are connected by two μ_3–O(1)H and six μ_3–O(2)H groups to form a homometal octahedral cluster $[Gd_6(OH)_8] \equiv [Gd_6]$ (Fig. 10b). The $[Gd_6]$ core is encapsulated by six symmetry-related $[Cu_2]$ fragments via the μ_3–O(3)H and another two glys, leading to the unique axial-fan-shaped cation $[Gd_6Cu_{12}]$ (Fig. 10c). Each $[Cu_2]$ fragment as the blade of the fan is fastened by three Gd nails onto the axial octahedral core. The two glycines in the blade employ coordination mode **d2** (as shown in Scheme 1) to chelate a Cu^{2+} ion and further to coordinate two Ln^{3+} ions via their two O atoms. The octa-aqua Gd^{3+} ion in the periphery is disordered, similar to that observed in *p*-sulfonatocalix [4]arene systems [101, 102].

Interestingly, compound **10** features the largest and "hollow" $[Gd_6]$ cluster without the support and "contracted" effect of a centered μ_6–oxo ligand. Except for the reported $[Tb_6]$ unit in the $[Tb_{14}]$ cluster [103–106], all previously known octahedral

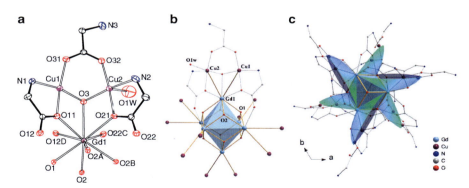

Fig. 10 a ORTEP view of **10** at 30% probability level shows the coordination geometries of Gd^{3+}, Cu^{2+} atoms. Hydrogen atoms, perchlorates, octa-aqua Gd^{3+} ion, and lattice water molecules are omitted for clarity. The symmetry codes for A, B, C, and D are $1-y$, $1+x-y,z$; $2/3+x-y$, $1/3+x$, $1/3-z$; $1/3+y$, $1/3-x+y$, $1/3-z$; and $-x+y$, $1-x,z$, respectively. **b** Six $[Cu_2]$ blades encapsulate an octahedral $[Gd_6]$ core to form a $[Gd_6Cu_{12}]$ cluster cation with an axial-fan shape viewed along the *c* axis as shown in **c**. In panel **b**, the $[Gd_6]$ octahedron is colored and the Cu atoms are labeled. The bridging angle of Cu1–O3–Cu2 is 105.91°. The torsion angles of Gd1–O3–O11–Cu1 and Gd1–O3–O21–Cu2 are 173.06° and 173.27° respectively

Ln$_6$ clusters in molecular compounds have a μ_6–oxo ligand in the center of the octahedron, which is believed to play a key role in stabilizing the Ln$_6$ unit [107–111]. The absence of μ_6–oxo ligand in the [Gd$_6$] core of **10** leads to the larger Gd–μ_3–OH and Gd–Gd distances (ranging from 2.389 to 2.416 and 3.955 to 3.959 Å), and the Gd–μ_3–O–Gd angles (110.16–111.60°), in comparison with the corresponding distances (2.345–2.408 and 3.5612–3.6204 Å) and angles (97.81–99.62°) in the reported [Gd$_6$] cluster perchlorate [109]. Actually, to the best of our knowledge, the mean Gd–μ_3–O–Gd angle of 110.92° in **10** is the largest among the polynuclear [Gd$_n$] oxo-clusters.

Temperature–dependent magnetic susceptibilities of **10** measured from ground crystalline samples in the temperature range of 2–300 K under various applied fields are shown in Fig. 11. The plot shows $\chi_M T$ vs. T per molecule, where χ_M is the molar magnetic susceptibility calculated with correction for the diamagnetic contribution of the compound. At room temperature (300 K) and a field of 500 Oe, the measured $\chi_M T$ value is 48.8 cm^3 mol^{-1} K, as compared with the expected value of 59.6 cm^3 mol^{-1} K for free and noninteracting 12 Cu^{2+} ions and seven Gd^{3+} ions. The difference may result from strong antiferromagnetic contribution within the six [Cu$_2$] blades of the compound. Upon cooling, **10** shows a continuous slight increase in $\chi_M T$, and subsequently a much sharper increase below ca. 40 K, with a maximum value of 52.5 cm^3 mol^{-1} K at 10 K, suggesting that **1** exhibits overall ferromagnetism, as confirmed by a positive Weiss constant (+0.8 K). Then, $\chi_M T$ drops rapidly to 42.8 cm^3 mol^{-1} K at 2 K, which is mainly attributed to weak intermolecular antiferromagnetic coupling and zero field splitting [112]. A similar situation is also observed at fields of 1 and 5 KOe. The maximum value of $\chi_M T$ at 10 K slightly increases with increasing magnetic field, and that of 53.8 cm^3 mol^{-1} K at 5 KOe is still smaller than the spin-only value, indicating the existence of antiferromagnetic interaction or spin frustration. However, by further increasing the field (10 and 25 KOe), the data display significant field-dependence, reaching a maximum that shifts to higher temperature. This behavior indicates the presence of a high-spin

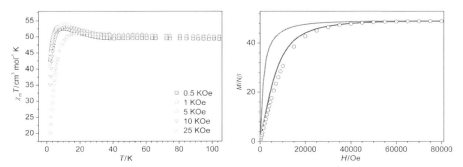

Fig. 11 *Left*: experimental $\chi_M T$ vs. T for **10** at H_{dc} of 0.5, 1, 5, 10, and 25 KOe. *Right*: field dependence of the magnetization of **10** at 2 K (*open circle*) compared with the Brillouin function for seven uncoupled Gd^{3+} ions (*blue line*) and the Brillouin function for an $S = 42/2$ state plus an $S = 7/2$ state (*red line*)

ground state [113, 114]. The decrease of the maximum value at high fields and low temperatures may be attributed to depopulation of the highest-energy Zeeman levels [113, 114], and to magnetic decoupling between Gd ions [115, 116].

From two empirical formulas suggested by the pioneers, the interaction for Cu–Cu coupling (J') with the Cu–O–Cu angle (ϕ) of 105.91° may be antiferromagnetic in nature with a magnitude of about $623.5\,cm^{-1}$ ($J'(cm^{-1}) = -74.53\phi + 7270$) [88], while that between Cu–Gd ions (J) with the dihedral angles (c) of 6.8° may be ferromagnetic with a magnitude of about $8.0\,cm^{-1}$ [$|J|(cm^{-1}) = 11.5\exp(-0.054c)$, where c is the dihedral angle between the two halves of the CuO_2Gd bridging unit] [117]. So, frustration will occur within the triangular $GdCu_2$ system (Fig. 10a). Indeed, the magnetization curve gives a large saturated value of $48.9\,N\beta$ that is close to the expected value of $49\,N\beta$ for seven Gd^{3+} ions (Fig. 11). In the high-field regime (≥ 25 KOe), single-ion behavior of Gd^{3+} is observed. This suggests that at high field, the two unpaired electrons on each of the six [Cu_2] blades will be forced to pair so that the spin frustration effect within the $GdCu_2$ triangle will be minimized. It is worth mentioning that at 25 KOe, the weak ferromagnetism for **10** still remains with a Weiss constant of $+0.5$ K. In the low-field regime, the magnetization values lie below the calculated curve for the isolated Gd^{3+} system, indicating that **10** is a ferrimagnet with strong spin frustration.

From the exchange and structural parameters of the reported complexes containing the GdO_2Gd unit, an empirical formula:

$$J'' = 0.0123\varphi - 1.364cm^{-1},\tag{2}$$

was obtained for the first time, where φ and J'' are the mean Gd–O–Gd bridging angle and the exchange parameters between Gd ions, respectively. It can be concluded that $\varphi > 110.9°$ is a prerequisite for a GdO_2Gd complex exhibiting molecular ferromagnetism. Therefore the "hollow" [Gd_6] core in compound **10** has the largest mean Gd–μ_3–O–Gd φ, which is slightly larger than the critical value above, suggesting a very weak ferromagnetic interaction with a magnitude of about $0.00035\,cm^{-1}$.

The magnetic properties of **10** are dominated by three types of Cu–Cu, Cu–Gd and Gd–Gd coupling. It is well known that the interaction sequence 4f–4f < 4f–3d < 3d–3d is important as far as the mechanism of the interaction phenomenon is concerned. At low field (500 Oe), the magnetic structure of compound **10** can be described as six spin-frustrated $GdCu_2$ triangles plus one isolated Gd^{3+} ion, while at high field (2.5 KOe), its magnetic structure can be seen as a Gd_6 octahedron. After fitting the magnetic data at the two fields above, the exchange constants for the Cu–Cu, Cu–Gd and Gd–Gd coupling can be obtained as -857.3, 9.6 and $0.0016\,cm^{-1}$, respectively. The positive exchange constants for Gd–Gd coupling from empirical estimation and data fitting suggest that the [Gd_6] core in **10** is the first example of a high-nuclearity [Ln_n] cluster exhibiting molecular ferromagnetism. The magnetic property of **10** is mainly predominated by antiferromagnetic Cu–Cu coupling and ferromagnetic Gd–Gd and of Gd–Cu coupling. At low field, compound **10** is a frustrated ferrimagnet, while it exhibits the single-ion behavior of Gd^{3+} ions at high field.

6 Effect of Lanthanide(III) Ions on the Assembly

Generally Ln^{3+} ions with larger ionic radii prefer high coordination numbers (sometimes up to 12), while low coordination numbers occur for those with smaller ionic radii. In general, this is also known as lanthanide contraction. In many cases, a series of isostructural compounds can be synthesized with lanthanide ions varying from La^{3+} to Lu^{3+}, as described above. However, when solutions containing Ln(ClO$_4$)$_3$, Cu(ClO$_4$)$_2$, and glycine in a molar ratio of 2:1:2 were left in the air, rather than placed in a desiccator filled with phosphorus pentaoxide, slow evaporation yielded three distinct structural types of 3d–4f compounds with various lanthanides. Firstly, the La, Pr, and Sm compounds are 1D catenated polymers {Ln(H$_2$O)$_3$[Cu(gly)$_2$][Cu(gly)$_2$(H$_2$O)]}$_2$ · 6ClO$_4$ · 4H$_2$O (**11**) composed of lanthanide dimers that are bridged by four blade-like *cis*-Cu(gly)$_2$ groups, in which its repeating unit may be described as a four-bladed propeller. Next, in the Eu (**12**) and Dy (**13**) compounds, 2D network are composed of [Ln$_6$Cu$_{22}$] clusters linked by *trans*-Cu(gly)$_2$ groups to form a porous material. Finally, a 3D Er compound (**14**) is made up of [Er$_6$Cu$_{24}$] clusters connected by *trans*-Cu(gly)$_2$ groups. Each [Er$_6$Cu$_{24}$] cluster interconnects eight symmetry-related {Er$_6$Cu$_{24}$} clusters through 12 *trans*-Cu(gly)$_2$ groups to form a 3D porous material with a number of ClO$_4^-$, Na$^+$, and lattice water molecules accommodated in the channels.

6.1 {Ln(H$_2$O)$_3$[Cu(gly)$_2$][Cu(gly)$_2$(H$_2$O)]}$_2$ · 6ClO$_4$ · 4H$_2$O(11) (Ln^{3+} = La, Pr, Sm)

The three compounds are isostructural and only the structure of **11·Sm** is described here. An ORTEP view of compound **11·Sm** is shown in Fig. 12. There exist one crystallographically independent Sm^{III} ions, two Cu^{II} ions, four glycinato ligands, three perchlorate ions, and four coordinating plus two noncoordinating water molecules. The Sm1 center is coordinated by nine oxygen atoms from five glycinato

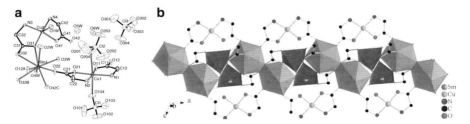

Fig. 12 a ORTEP view of compound **11·Sm** with 50% thermal ellipsoids. All hydrogen atoms are omitted for clarity. The symmetry codes for A, B, and C are −1 + x, y, z; −2 − x, −y, 1 − z; −1 − x, −y, 1 − z, respectively. **b** Stereo view of the chain-like polymer of **11·Sm**

anions and three coordinating water molecules. The coordination geometry of the Cu1 center can be described as a distorted octahedron with [4 + 2] mode. The basal plane of the octahedron is occupied by two glycinato chelators with average Cu1–O and Cu1–N distances of 1.938 and 1.991 Å, respectively, whereas the axial positions are occupied by perchlorate oxygen atoms with long average Cu1–O distances 2.592 Å. The basal plane [Cu(gly)$_2$] is planar to within ±0.110 Å. The Cu2 ion is also chelated by two glycinato ligands to form a distorted plane with mean deviation of 0.232 Å, and its square pyramidal configuration is completed by a water ligand at the apical position (Cu2–O1w distance is 2.357 Å). The dihedral angle between the basal planes of two Cu ions is 54.3°. The Cu–N and Cu–O distances in **11** range from 1.988 to 2.006 Å and from 1.934 to 2.611 Å, respectively, which are similar to those found in the related amino acid–copper species [53–56]. The four glycinato ligands in **11·Sm** adopt two coordination modes, **c0** and **1e** as depicted in Scheme 1. All glycinato ligands are deprotonated, in agreement with the IR data in which no strong absorption peaks around 1700 cm^{-1} for –COOH are observed.

As shown in Fig. 12b, the Sm1, Cu1, and Cu2 ions are connected together via four glycinato ligands to generate a chain along the *a* axis. Two symmetry-related Sm ions connect with each other through two carboxylate oxygen atoms to form an edge-shared dimer. Four blade-like diglycinato–copper groups bridge the dimers of two neighboring barrel units, generating a catenated chain in which the repeating unit may be described as a four-blade propeller. Three perchlorate ions with large thermal parameters and two non coordinating water molecules lie between the propellers. Pairs of O5w and O1w molecules of two neighboring chains are related by an inversion center to give a rectangle held by hydrogen bonds, which connects the chains to form a layer along the *ab* plane. However, there is no strong interaction between adjacent layers other than Van der Waals forces.

Compounds **11·La** and **11·Pr** are isostructural with **11·Sm**. Nevertheless, small difference exists for these three compounds. As the ionic radii of the lanthanides decreases from 1.17, 1.13 to 1.10 Å, the size of the Ln$_2$ nodes more distinctly decreases, with the *x* value varying from 6.000, 5.878 to 5.775 Å and the Y value from 4.989, 4.891 to 4.804 Å, while the angle (θ) between the *a* axial and the connection line of the Ln$_2$ dimer also decreases from 144.1, 144.0 to 143.9° (Table 1). The Ln$_2$ dimer's size shrinks even more remarkably than the cell parameters, which makes the two metalloligands highly disordered, especially for the Cu(1)(gly)$_2$(H$_2$O) ligands linking Ln1 and Ln1A to form the dimers and simultaneously connecting adjacent dimers to form the chain whose torsion angle (TA) between the two halves increases from 26.5, 27.5 to 28.1°. This is the first observation of a large torsion angle for the Cu(gly)$_2$ ligand, while the corresponding angles for the cluster {Na $\subset$ Cu$_2$[Cu(gly)$_2$]$_4$} range from 5.4 to 9.3° [53], and that for the compound Cu(gly)$_2$(H$_2$O) is only 4.0° [118].

Controllable Assembly, Structures, and Properties of Lanthanide–Transition 187

Table 1 Some unit cell parameters, distances (Å) and angles (°) for compounds **11**

			11 · La	**11 · Pr**	**11 · Sm**
Unit cell parameter	a		9.7475(11)	9.6817(50)	9.6132(19)
	b		11.9945(18)	11.9617(67)	11.9083(24)
Ln2	c		13.6087(23)	13.6023(76)	13.5959(27)
	x^a		6.0002	5.8782	5.7746
	y^a		4.9890	4.8909	4.8045
	θ^a		144.1°	144.0°	143.9°
CU2	Cu2–N3[b]		1.985 ± 0.0657	1.991 ± 0.0750	2.006 ± 0.0773
	Cu2–N4[b]		1.979 ± 0.0826	1.992 ± 0.0821	1.999 ± 0.0906
	TA[c]		26.5°	27.5°	28.1°
CU1	Cu1–N2[b]		1.997 ± 0.0364	1.984 ± 0.0429	1.988 ± 0.0491
	Cu1–N1[b]		1.990 ± 0.0524	1.987 ± 0.0564	1.994 ± 0.0521
	TA[c]		9.3°	10.6°	12.0°

[a] x and y (Å) and angle θ (°) are indicated in the figure
[b] Cu–N distances (Å) and the mean deviation of the plane composed of the corresponding Cu and gly
[c] TA is the torsion angle (°) of the two Cu(gly) planes

6.2 $Na_2[Eu_6Cu_{22}(OH)_{28}(Hgly)_4(gly)_{12}(ClO_4)(H_2O)_{18}][Cu(gly)_2]_3 \cdot 23ClO_4 \cdot 28H_2O(12)$ and $[Dy_6Cu_{22}(OH)_{28}(Hgly)_4(gly)_{12}(ClO_4)(H_2O)_{18}][Cu(gly)_2]_3 \cdot 21ClO_4 \cdot 20H_2O(13)$

When the lanthanides are Eu^{3+} or Dy^{3+} ions, blue crystals Na_2 $[Eu_6Cu_{22}(OH)_{28}(Hgly)_4(gly)_{12}(ClO_4)(H_2O)_{18}][Cu(gly)_2]_3 \cdot 23ClO_4 \cdot 28H_2O$ **(12)** and $[Dy_6Cu_{22}(OH)_{28}$ $(Hgly)_4(gly)_{12}(ClO_4)(H_2O)_{18}][Cu(gly)_2]_3 \cdot 21ClO_4 \cdot 20H_2O$ **(13)** can be separated from the parent solution with the Ln:Cu:gly ratio of 2:1:2 at pH value of 6.6. The most significant feature of **12** and **13** is a 2D network composed of defective $[Ln_6Cu_{22}]$ cluster nodes and *trans*-Cu(gly)$_2$ linkers, as shown in Figs. 13 and 14.

There are three crystallographically independent Eu^{3+} ions, 13 Cu^{2+} ions and 11 glycinato ligands in compound **12 · Eu**. For the 11 glycine ligands, five of them are coordinated in mode **d2**, three in mode **c1**, two in mode **a**, and one in mode **c2** (see Scheme 1). The **c2** mode is for the first time observed for amino acid ligands in 3d–4f compounds. Each Eu^{3+} here is coordinated by nine oxygen atoms with the coordination polyhedron described as a monocapped square antiprism. As shown in Fig. 13a–c, both Eu1 and Eu3 ions coordinate to four inner μ_3–OH$^-$ groups (lower plane), two carboxylate oxygen atoms, two water molecules (upper plane), and one outer μ_3–OH$^-$ group (cap), while Eu2 ion has a similar coordination environment other than its cap, which is occupied by the O71 atom from another glycine ligand. The 13 Cu^{2+} ions can be divided into three kinds of coordination geometries: (i) six inner Cu^{II} centers each coordinated by six oxygen atoms in an octahedron; (ii) five outer Cu^{II} centers exhibiting obvious Jahn–Teller distortion; and (iii) two bridging Cu^{II} centers in square planar coordination mode, which are each

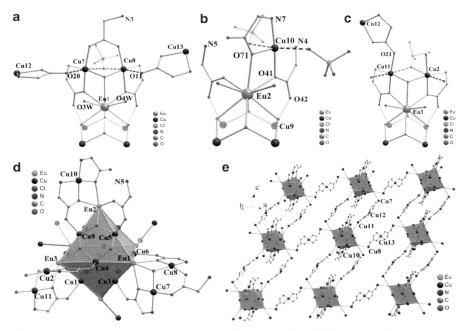

Fig. 13 a–c Coordination mode of the outer Cu^{2+} ions and the Eu^{3+} ions at the three vertices of the huge octahedral cluster for **12**. **d** [Er_6Cu_{22}] cluster nodes. **e** 3D open-framework of 12 viewed along [−1 − 10] direction

linked by two glycines through an amino nitrogen atom and a carboxylato oxygen to form the *trans*-Cu(gly)$_2$ group.

The 12 inner Cu^{II} ions and 6 Eu^{3+} ions are connected by 24 inner μ_3–OH$^-$ groups to form a [Eu_6Cu_{12}] core with the help of a μ_{12}–ClO$_4^-$ groups, similar to the core in the triacontanuclear 3d–4f amino acid cluster compounds and the [Ln_6Cu_{12}] clusters with carboxylate ligands [81–85]. The five crystallographically independent outer Cu^{2+} ions are six-coordinated with the [4 + 2] mode. Among them, both Cu7 and Cu8 ions have an NO$_5$ donor set that consists of one amino nitrogen atom and one carboxylate atom from a glycine in the η_2-mode, one outer μ_3–OH$^-$, one carboxylate oxygen from one *trans*-Cu(gly)$_2$ bridge (the distances of Cu7–O20 and Cu8–O11 are 2.3064 and 2.3041 Å respectively), and one oxygen atom from a perchlorate ion. The O20 and O11 atoms from the *trans*-Cu(gly)$_2$ group occupy the sixth coordinated sites of Cu7 and Cu8 ions, respectively. The NO$_5$ donor set for Cu2 and Cu11 ions is similar to that for Cu7 ion except that the sixth coordinated sites of Cu9 and Cu11 ions are occupied by water molecules, and their fourth coordinated sites are respectively occupied by a water molecule and a carboxylate oxygen from the *trans*-Cu(12)(gly)$_2$ group. The Cu10 ion is chelated by two glycine ligands to form a *cis*-Cu(gly)$_2$ group, which binds to one Eu2 ion via O41 and O71 atoms with Eu–O distances of 2.5255 and 2.5133 Å, respectively, and to one inner Cu^{2+} ion via another carboxylate oxygen atom (Cu9–O42 2.2780 Å). Then,

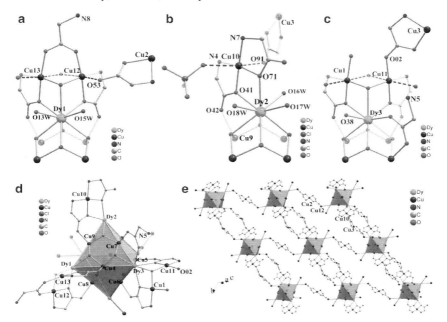

Fig. 14 a–c Coordination mode of the outer Cu^{2+} ions and Dy^{3+} ions at the three vertices of the huge octahedral cluster for **13**. **d** [Dy$_6$Cu$_{22}$] cluster nodes. **e** 3D open-framework of **13** viewed along [1 0 −1] direction

the Eu2 ion only connects one outer Cu^{2+} ion through two glycines in modes **c2** and **d1**, while Eu1 and Eu3 ions respectively link two outer Cu^{2+} ions through one outer μ$_3$–OH$^−$ group and two glycine ligands, both in mode **d1**. Thus, an octacosanuclear [Eu$_6$Cu$_{22}$] cluster is formed. Compared with the triacontanuclearity clusters mentioned above, the [Eu$_6$Cu$_{22}$] cluster nodes in compound **12** can be considered as defective clusters. The defect results from a *cis*-Cu(10)(gly)$_2$ group that chelates the Eu2 ion via its two chelating carboxylate oxygen atoms. Chelation of the *cis*-Cu(10)(gly)$_2$ group to Eu2 ion hinders the formation of a binuclear copper-gly fragment [Cu$_2$(gly)$_3$(OH)(H$_2$O)] ≡ [Cu$_2$] in the two Eu2 vertices of the large octahedron cluster. In other words, the ratio 1:2 of Cu^{2+} ion to glycine favors formation of the Cu(gly)$_2$ group, which further leads to generation of the defective [Eu$_6$Cu$_{22}$] cluster node.

Besides the *cis*-Cu(gly)$_2$ group within the [Eu$_6$Cu$_{22}$] cluster node, there are two other Cu(gly)$_2$ groups in *trans*-mode, which link the [Eu$_6$Cu$_{22}$] nodes to form the 2D framework. The *trans*-Cu(12)(gly)$_2$ group employs its oxygen atoms O20 and O21 to respectively occupy the six-coordinated site of Cu7 ion and the fourth coordinated site of Cu11 ion in an adjacent [Eu$_6$Cu$_{22}$] cluster in the next cell (Cu11–O21 1.9759 Å). The Cu13 ion is located at the 1*e* position, and its *trans*-Cu(gly)$_2$ group employs two crystalographically dependent O11 atoms to occupy the six-coordinated sites of two Cu8 ions in adjacent clusters. In the crystal, each [Er$_6$Cu$_{22}$] unit is firstly polymerized through two *trans*-Cu(12)(gly)$_2$ bridges to yield a 1D double-chain running parallel to the [1] direction. Then these double-chains are further connected by the *trans*-Cu(13)(gly)$_2$ group. That is, each [Eu$_6$Cu$_{22}$] unit is connected through six *trans*-Cu(gly)$_2$ bridges to four neighbors, resulting in a net-like structure extended along the [001] and [1–10] directions (Fig. 13d).

Similar to compound **12**, compound **13** also comprises defect [Dy$_6$Cu$_{22}$] nodes, as shown in Fig. 13. Each node is connected through six *trans*-Cu(gly)$_2$ to four neighboring [Dy$_6$Cu$_{22}$] units, resulting in a 4^4-net structure extended along the [111] and [101] directions. But, because of the lanthanide contraction effect, there are several differences, including their nodes and the connection mode, between the *trans*-Cu(gly)$_2$ bridges and the nodes. In the two compounds, the *cis*-Cu(gly)$_2$ group must chelate one Ln ion and synchronously bind to another Cu^{2+} ion from the [Ln$_6$Cu$_{12}$] core. Furthermore, its two chelating O atoms must still accommodate the monocapped-square-antiprismatic coordination environments of the Ln ion. Therefore, when the lanthanide changes from Eu to Dy with the corresponding change in ionic radii from 1.12 to 1.083 Å, the barycenter of the *cis*-Cu(gly)$_2$ group will shift to the direction of the vertex (Ln ion) of the [Ln$_6$Cu$_{12}$] octahedron, accompanied by the shortened Dy2–O71 distance (2.4693 Å) and lengthened Cu9–O42 one (2.2943 Å). Firstly, the shift makes the *cis*-Cu(gly)$_2$ group occupy the more outer space of the vertex (Ln ion) of the octahedron, which then hinders the binding of the glycinato ligand (N5) to the same Dy2 ion, while the glycinato ligand (N5) binds to the same Eu2 ion together with the *cis*-Cu(gly)$_2$ in compound **12**. In order to complete the O$_9$ coordination environment of the lanthanide, each the Eu^{3+} ion in compound **12** is coordinated by two water molecules. For the three crystallographically independent Dy^{3+} ions in compound **13**, Dy1 ion is also coordinated by two, Dy3 by one, and Dy2 by three water molecules. Secondly, the shift creates more space in the neighborhood of the Cu10 ions so that they can be linked by larger linkers (the *trans*-Cu(gly)$_2$ groups in compound **13**) while the two apical sites of the Cu10 ions are occupied by the O atoms from two perchlorate ions in compound **12**. Finally, the shift also lengthens the Cu10–N7 distance from 1.9804 to 2.0346 Å and the Cu10–N4 distance from 1.9802 to 2.007 Å. To date, the longest Cu–N bond length is 2.021 Å for all compounds containing the Cu(gly)$_2$ fragment in the Cambridge Structural Database [118]. The Cu10–N7 bond length in compound **13** may be the limiting value for the Cu(gly)$_2$ compounds. Therefore, the *cis*-Cu(gly)$_2$ group tends not to coordinate to the [Ln$_6$Cu$_{12}$] octahedron for lanthanide ions with sizes smaller than that of DY^{3+}.

6.3 *Na$_4$[Er$_6$Cu$_{24}$(Hgly)$_2$(gly)$_{12}$(OH)$_{30}$(ClO$_4$)(H$_2$O)$_{22}$][Cu(gly)$_2$]$_6$ · 27ClO$_4$ · 36H$_2$O (14)*

When the lanthanide used is the Er^{3+} ion, blue crystals Na$_4$[Er$_6$Cu$_{24}$ (Hgly)$_2$(gly)$_{12}$ (OH)$_{30}$(ClO$_4$)(H$_2$O)$_{22}$]· [Cu(gly)$_2$]$_6$ · 27ClO$_4$ · 36H$_2$O (**14**) can be separated from the parent solution, with the molar ratio Er:Cu:gly set to 2:1:2 at a pH of 6.6. The most interesting feature of **14·Er** is a 3D network comprising [Eu$_6$Cu$_{24}$] cluster nodes and *trans*-Cu(gly)$_2$ group linkers, as shown in Fig. 15.

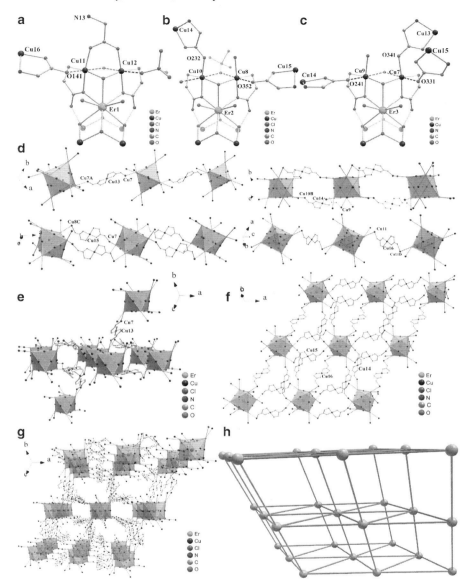

Fig. 15 a–c Coordination mode of the outer Cu^{2+} ions and the Er^{3+} ions at the three vertices of the cluster $[Er_6Cu_{24}]$ for **14**. **d** $[Er_6Cu_{24}]$ cluster nodes linked by the four *trans*-Cu(gly)$_2$ groups to generate the chains along [110], [011], [100], and [111] directions, respectively. Symmetry codes for A, B, C, and D are $-1-x, -1-y, -z$; $x, 1+y, 1+z$; $-1+x, y, z$; $1-x, 1-y, 1-z$, respectively. **e** Each cluster node links to eight cluster units through 12 *trans*-Cu(gly)$_2$ groups. **f** 2D sheet parallel to the [111] plane. **g** 3D open-framework viewed along the [111] direction. **h** Network of $3^6 4^{18} 5^3 6$ topology

There are three crystallographically independent Er^{3+} ions, 16 Cu^{2+} ions, and 13 glycinato ligands in compound $14 \cdot Er$. For the 13 glycine ligands, six of them are coordinated in mode **d2**, six in mode **c1**, and one in mode **a** (see Scheme 1 for representations of the modes). The glycines in modes **d2** and **a** are used to construct the $[Eu_6Cu_{24}]$ cluster nodes, while the ones in mode **c1** chelate Cu ions to form the *trans*-Cu(gly)$_2$ group linkers. Each Er^{3+} here is coordinated by nine oxygen atoms in the configuration of a monocapped square antiprism. The nine oxygen atoms are composed of four inner μ_3–OH$^-$ groups (lower plane), two carboxylate oxygen atoms, two water molecules (upper plane), and one outer μ_3–OH$^-$ group (cap). The Er–O bond lengths are in the range 2.370–2.504 Å.

The 16 Cu^{2+} ions can be divided into three kinds of coordination geometries: (i) six inner Cu^{II} centers each coordinated by six oxygen atoms in an octahedron; (ii) six outer Cu^{II} centers exhibiting obvious Jahn–Teller distortion; and (iii) four bridging Cu^{II} centers in square–planar coordination mode, which are each linked by two glycines through an amino nitrogen atom and a carboxylato oxygen to form the *trans*-Cu(gly)$_2$ group. The 12 inner Cu^{II} ions and six Er^{3+} ions are connected by 24 inner μ_3–OH$^-$ groups to form a $[Er_6Cu_{12}]$ core with the help of a μ_{12}–ClO$_4^-$ group, similar to the core in the triacontanuclear 3d–4f-amino acid cluster compounds and the $[Ln_6Cu_{12}]$ clusters with carboxylate as ligands [81–85]. The six crystallographically independent outer six-coordinated Cu^{2+} ions have an NO$_5$ donor set with $[4+2]$ mode. The first three coordinated sites for these outer Cu^{2+} ions are all occupied by an amino group and a carboxylate oxygen from a $[3.2_{12}2_{23}1_3]$-mode glycine, and a μ_3–OH group. A $[2.110]$-coordinated glycine provides its two oxygen atoms to respectively occupy the fourth sites of Cu11 and Cu12 ions, whereas the fourth sites of Cu7 and Cu10 ions are occupied by oxygen atoms from the *trans*-Cu(gly)$_2$ group (Cu7–O341 1.9876 Å; Cu10–O232 1.9632 Å), and those of Cu8 and Cu9 ions by water molecules. The sixth coordinated sites of Cu7, Cu8, Cu9, and Cu11 ions are occupied by oxygen atoms from *trans*-Cu(gly)$_2$ groups with Cu–O distances of 2.3656, 2.3220, 2.4110, and 2.2688 Å. Among the six outer Cu ions of the $\{Er_6Cu_{24}\}$ cluster node, Cu7 ion is linked by two *trans*-Cu(gly)$_2$ groups, which has not been observed in previously reported Cu–Ln amino acid cluster compounds.

Among the four *trans*-Cu(gly)$_2$ groups, Cu(13) is located at the $1e$ position, and the *trans*-Cu(13)(gly)$_2$ group uses its two carboxylate oxygen atoms to occupy the fourth sites of two crystallographically dependent Cu7 ions of adjacent clusters to form a chain along the [110] direction, which was also observed for the first time in Cu–Ln amino acids systems. Cu(16) resides at the $1h$ position, and its *trans*-Cu(gly)$_2$ group links adjacent clusters to form a chain along the [100] direction by occupying the sixth sites of the two crystallographically dependent Cu11 ions. The *trans*-Cu(14)(gly)$_2$ group employs its two oxygen atoms O241 and O232 to respectively occupy the six-coordinated site of Cu9 ion and the fourth coordinated site of Cu10 ion of the adjacent $[Er_6Cu_{24}]$ cluster in the next cell, thus forming a double-chain along the [011] direction. Similarly, the *trans*-Cu(15)(gly)$_2$ group links adjacent cluster nodes to form a double-chain along the [100] direction. The $[Er_6Cu_{24}]$ cluster nodes are linked by three *trans*-Cu(gly)$_2$ groups (Cu14, Cu15, and Cu16) to form a layer extended along the [011] and [100] directions. The *trans*-Cu(13)(gly)$_2$ groups link adjacent layers to give a 3D open-framework, which contains 1D rectangular channels with dimensions of about 10.1×20.0 Å (based on the Cu7$\cdots$Cu7 and Cu13$\cdots$C̄u13 separations) along the [111] direction. The effective free volume of **14** is about 3583.5 Å^3, comprising 53.9% of the crystal volume, as calculated by the program PLATON. Free water molecules, perchlorate ions and sodium cations are encapsulated in the large pores.

Metal–organic frameworks (MOFs) based on nets with coordination numbers ≥ 8 are rare [119]. In our previous reports, the eight-connected compound **8** has each [Sm$_6$Cu$_{24}$] cluster node linked with six adjacent cluster nodes through ten *trans*-Cu(gly)$_2$ groups in a primitive cubic net structure with Schläfli symbol $4^{12}6^3$; and the 12-connected compound **9** has each [Nd$_6$Cu$_{24}$] cluster node linked with 12 adjacent nodes though 12 *trans*-Cu(pro)$_2$ groups to form a face-centered cubic network with Schläfli symbol $3^{24}4^{18}5^36$ [95]. In contrast, in compound **14**, each {Er$_6$Cu$_{24}$} cluster node connects eight adjacent ones through 12 *trans*-Cu(gly)$_2$ groups to form a 3D uninodal eight-connected framework of $3^64^{36}5^36^3$ topology. Since both Cu(14)(gly)$_2$ and Cu(15)(gly)$_2$ linkers connect the [Er$_6$Cu$_{24}$] nodes to form the double-chain, the ten *trans*-Cu(gly)$_2$ groups (four Cu14, four Cu15, and two Cu16) link [Er$_6$Cu$_{24}$] nodes into an independent series of parallel 3^6-nets. At the same time, a series of parallel 4^4-nets can be obtained through the linkage of *trans*-Cu(13)(gly)$_2$ groups and one of the other three *trans*-Cu(gly)$_2$ groups. Thus, this lattice can be viewed most easily as comprising intersecting 4^4- and 3^6-nets. Such 3^6-net layers extended along the [011] and [100] directions are also observed in compound **9**, each of which is further linked by three pairs of *trans*-Cu(pro)$_2$ linkers to give the face-centered cubic network. In compound **8**, the layer parallel to the [111] plane is a 4^4-net, which intersects with another 4^4-net to produce the primitive cubic net structure. In other words, the assembly of high-nuclearity clusters with amino acids as ligands is an effective means for the construction of MOFs characterized by high coordination number and interesting network topology.

7 Constitutionally Dynamic Chemistry of Cu–Ln–gly Compounds

The isoelectric point for glycine is 5.9. Under the reaction conditions at a pH of 6.6, the zwitterionic (Hgly) or anionic (gly) forms of glycine coexist in equilibrium. As shown in Scheme 2, if Cu^{2+} ions are added into the solution, the anionic gly ligands may chelate Cu ions to form a [Cu(gly)(H$_2$O)$_2$]$^+$ metalloligand (*a*), two of which can be linked by one OH$^-$ group to form another metalloligand [Cu$_2$(gly)$_2$(OH)(H$_2$O)$_2$]$^+$ (*b*). Like the H$_2$O molecules in *a*, the two molecules in *b* also can be replaced by zwitterionic Hgly to form another metalloligand [Cu$_2$(gly)$_2$(OH)(Hgly)]$^+$ (*c*). On the other hand, *a* also can be further chelated by

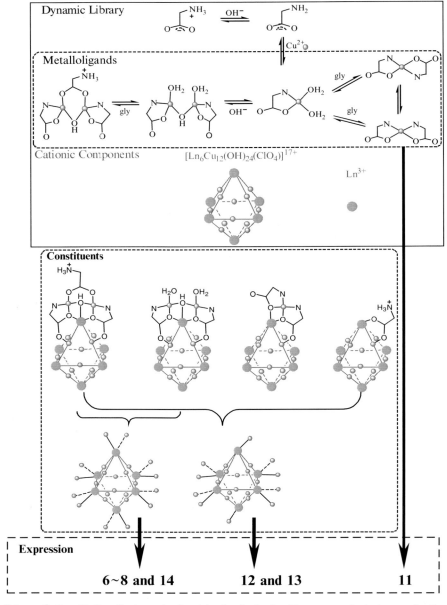

Scheme 2 Constitutionally dynamic chemistry for the lanthanide–copper–glycinato coordination polymers, showing metalloligands *a–f*

another gly to generate Cu(gly)$_2$ in either the *cis* (*d*) or *trans* (*e*) mode. Interconversion between the *cis*- and *trans*- Cu(gly)$_2$ isomers can be rapidly realized in aqueous solution, and even in the solid state under thermally driven conditions [120]. The

Controllable Assembly, Structures, and Properties of Lanthanide–Transition 195

metalloligands b–e are observed in compounds **6**–**14**, while a is also observed in a 32-nuclearity $Na_4[Tb_6Cu_{26}(\mu_3–OH)_{30}(gly)_{18}(ClO_4)–(H_2O)_{22}]\cdot(ClO_4)_{25}\cdot(H_2O)_{42}$ cluster (**15**) [92].

Similar to the glycinato ligand, the OH^- group can bridge the Cu^{2+} and Ln^{3+} ions to form an octahedral $[Ln_6Cu_{12}(OH)_{24}(ClO_4)]^{17+}$ cluster constituent (f) with the help of a perchlorate as the template at the center of the cluster. The constituent f is stabilized by several carboxylates [81–85]. Note that metalloligands b and c prefer the octahedral component f to single lanthanides, because the torsion angle between the two Cu(gly) fragments in both b and c is near $90°$. Taking compounds **6**–**8** as examples, the torsion angles for b (Fig. 6a) and c (Fig. 6c) in **6** are 87.1 and $98.3°$, respectively, while those for c (Fig. 7a, 8c) in **7·Er** and **8** are respectively 91.6 and $82.2°$, depending on the number of $trans$-$Cu(gly)_2$ that provide their O atoms to occupy the fifth coordination site of the Cu ions of metalloligand c. The two water molecules in metalloligand b can be replaced by O atoms from the $trans$-$Cu(gly)_2$ groups. In this case, the torsion angles would have a large difference, ranging from 73.4 (Fig. 7b) to $99.3°$ (Fig. 8b). Other than b and c, d and Hgly can also chelate the octahedral f to form several intermediates, which can be further chelated by these ligands to generate $[Ln_6Cu_{24}]$ cluster nodes if all the six vertices of the octahedron are chelated by b and/or c, or the defect $[Ln_6Cu_{22}]$ nodes if two of the six are chelated by d. The $trans$-$Cu(gly)_2$ groups bridge the $[Ln_6Cu_{24}]$ nodes to form 1D **6**, 2D **7**, and 3D **8** and **14**; in contrast, bridging the $[Ln_6Cu_{22}]$ nodes forms 2D **12** and **13**.

It is worthy of note that the pentadentate metalloligands b and c can chelate the octahedral f more tightly than the tridentate cis-$Cu(gly)_2$ group (d) and bidentate Hgly. Therefore, only when there are a number of $Cu(gly)_2$ in the parent solution, can chelation of d to octahedral f take place. From the molecular formulas of a–e, one can see that the ratio 1:1 of Cu:gly will favor the formation of a and b. Similarly, the ratio 1:1.5 leads to c, and 1:2 to d and e. In the case of 1:2, the cis-$Cu(gly)_2$ groups can directly bind to Ln^{3+} ions to form compound **11** if the Ln ion is big enough (Ln = La–Sm). Comparing the reactant ratios and the molecular structures for compounds **7** [94] and **11**–**14**, we can also conclude that the ratio 1:2 of Cu^{2+}:gly favors the formation of $Cu(gly)_2$ groups. First, only the cis-$Cu(gly)_2$ group is observed in **11**. Second, in **12** and **13**, the defective cluster $[Ln_6Cu_{22}]$ nodes are formed due to the chelating cis-$Cu(gly)_2$ groups. Finally, **14** can be considered as constructed from **7·Er**, in which three more $trans$-$Cu(gly)_2$ groups insert into the 4^4-nets of **7·Er** and link the $[Er_6Cu_{24}]$ nodes to form the 3D open-framework. Conversely, in the preparation of compounds **6**–**8**, the ratios of Cu to gly are 3:2, 2:2, or 6:4 respectively, which favors the production of metalloligands a or b. So, the effect of lanthanide contraction does not play a significant role.

Similarly to the $[Ln_6Cu_{24}]$ clusters, compound **10** can be obtained through the chelation of metalloligand c to an octahedral $[Gd_6(\mu_3–OH)_8]^{10+}$ core via a stepwise growth process, as shown in Scheme 3a [100]. The acetate anion, similarly to Hgly, can replace two water molecules of b to form metalloligand $[Cu_2(gly)(OH)(Ac)]$ (g), an analog of c, six of which encapsulate f to form the discrete $[Ln_6Cu_{24}]$ cluster (**2**) [80]. Interestingly, if similar replacement of water by imidazole to produce new metalloligand $[Cu_2(gly)(OH)(im)_2]^{2+}$ (h) occurs, due to the strong repulsion

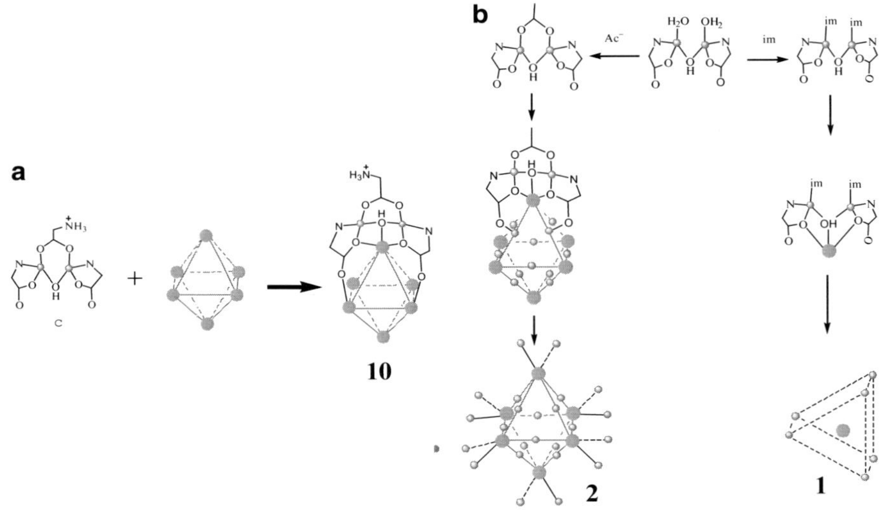

Scheme 3 Formation of compounds **a 10**, and **b 2** and **1**, showing metalloligands *g* and *h*

interaction between the two face-to-face imidazole rings, the distance of the two Cu ions will increase from 3.284 (*b* in **6**) to 3.497 Å (*h* in **1·La**), and the torsion angle of the two Cu(gly) fragments correspondingly decrease from 87.1 to 70.4°. Therefore, metalloligand *h* cannot chelate to octahedral *f* and only uses its three O donors to chelate Ln ions to form the heptanuclear trigonal prismatic cluster **1** [79].

The spontaneous aggregation of small building blocks in solution through multiple molecular recognition has been proven an effective way of constructing fascinating frameworks. In these systems, the fourth and fifth coordinated sites of the outer Cu^{2+} of the $[Ln_6Cu_{24}]$ nodes can be regarded as "recognition sites." The versatile coordination modes of Cu^{2+} ions due to Jahn–Teller distortion does benefit the constitutionally dynamic chemistry of Cu–Ln–gly compounds.

7.1 $Na_4[Tb_6Cu_{24}(OH)_{30}(gly)_{16}(ClO_4)(H_2O)_{18}]$ $[Cu(gly)(H_2O)_2]_2 \cdot 25ClO_4 \cdot 42H_2O(15)$

Compound **15** can be obtained through self-assembly from a parent solution containing $Tb(ClO_4)_3$, $Cu(ClO_4)_2$ and gly in a molar ratio of 1:2:1. The structure of its cation is shown in Fig. 16. The structure of **15** may be described as a $[Tb_6Cu_{24}]$ main structure connected with two $[Cu(gly)(H_2O)_2]^+$ groups. The $[Tb_6Cu_{24}]$ main structure is almost same as that in compound **2** [80] except for a slight difference in the 12 outer Cu^{2+} ions and the ligand. For the 12 outer Cu^{2+} ions, eight are four-coordinated in square–planar geometry. The other four adopt five-coordinated NO_4 square–pyramidal geometry. Two $[Cu(gly)(H_2O)_2]^+$ groups are connected to the

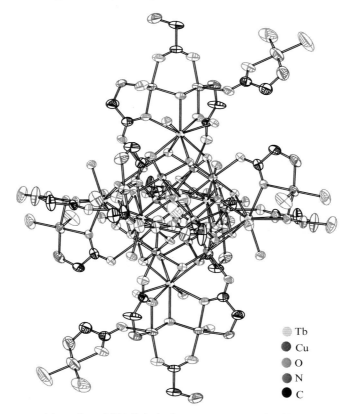

Fig. 16 Structure of the cation of **15** (all the hydrogen atoms are omitted for clarity)

[Tb$_6$Cu$_{24}$] main structure with the help of two glycinato ligands in mode **c0**. The terminal Cu^{2+} ion is coordinated by two water molecules, one nitrogen atom, and one oxygen atom from a glycinato ligand. The spare carboxylate oxygen atom of the gly ligand occupies the fifth coordinated site of an outer Cu^{2+} ion (Cu12), and thus [Cu(gly)(H$_2$O)$_2$]$^+$ groups and the [Tb$_6$Cu$_{24}$] main body are connected.

8 Effect of Transition Metal Ions on the Assembly

The versatile binding modes of the Cu^{2+} ion with coordination number from four to six due to Jahn–Teller distortion is one of the important reasons for the diverse structures of the Cu–Ln amino acid complexes. In contrast, other transition metal ions prefer the octahedral mode. For the divalent ions Co^{2+}, Ni^{2+}, and Zn^{2+}, only two distinct structures were observed: one is a heptanuclear octahedral [LnM$_6$] cluster compound, and the other is also heptanuclear but with a trigonal–prismatic structure.

For this type of octahedral clusters, the selection of the lanthanide ion is very important and only La^{3+} and Pr^{3+} compounds can be obtained [121]. For any particular lanthanide, the stoichiometry of the starting materials is crucial to generation of the final product. A Pr^{3+} : Ni^{2+} :gly molar ratio of 1:6:12 leads to the octahedral cluster in Na[PrNi$_6$(gly)$_{12}$](ClO$_4$)$_4$ · 11H$_2$O (**16**) [121], while the corresponding ratio 1:6:9 yields the trigonal–prismatic cluster in Na$_4$[PrNi$_6$(gly)$_9$(μ_3–OH)$_3$(H$_2$O)$_6$] · (ClO$_4$)$_7$ (**17**) [122]. It is worthy of note that the trigonal–prismatic structure is only obtained in systems with glycine as ligand, while the octahedral structure can occur in systems with gly, thr, ala, or pro as the ligand.

8.1 [LnM$_6$(AA)$_{12}$]$^{3+}$ Octahedral Clusters (M = Co^{2+}, Ni^{2+}; AA = gly, ala, thr)

The self-assembly of M(ClO$_4$)$_2$ (M = Ni, Co), Ln(ClO$_4$)$_3$ (Ln = La, Pr) and amino acids (glycine, L-alanine, and L-threonine) in aqueous media provides several heptanuclear octahedral clusters ([LnM$_6$(AA)$_{12}$]$^{3+}$) (AA = gly, ala, or thr) [121, 123]. The structure of a [LaNi$_6$] cluster (Na[LaNi$_6$(gly)$_{12}$](ClO$_4$)$_4$(H$_2$O)$_{11}$, (**16**) is shown in Fig. 17 as an example. The cationic structure is similar to the prolinato clusters reported by Yukawa [66], but the crystals are very stable in air. The La^{3+} ion is situated at the center of a large octahedral cage formed by six Ni^{2+} ions with La···Ni distances and Ni···Ni distances of about 3.6950–3.7019 and 5.2309–5.2424 Å, respectively, as compared with those observed in the [SmNi$_6$(pro)$_{12}$] cluster (3.7 and 5.23 Å, respectively) [66] and the [LaNi$_6$] cluster with iminocarboxylate as the ligand [124] (3.63 and 5.12–5.15 Å, respectively).

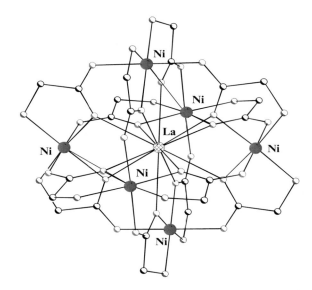

Fig. 17 Structure of the cation of **16**

Controllable Assembly, Structures, and Properties of Lanthanide–Transition 199

The O_{12} coordination polyhedron around the centered La^{3+} ion may be best described as an icosahedron. The 12 carboxyl oxygen atoms from 12 glycinato ligands coordinate to the central lanthanide ion with distances of about 2.692–2.761 Å. Each Ni atom has an N_2O_4 donor set that consists of two amino nitrogen atoms and four carboxyl oxygen atoms from four glycinato ligands. The Ni–O and Ni–N distances are about 2.04–2.07 Å. Slightly distorted octahedral configurations are found for all Ni ions. In the cation, each glycinato ligand adopts a μ_4-coordination mode, being bound to two Ni^{2+} and one La^{3+} ion.

An interesting phenomenon controlled mainly by the ionic radii of lanthanide ions was also observed. Only $[LaNi_6]$ and $[PrNi_6]$ octahedral clusters could be obtained though we have tried La^{3+}, Pr^{3+}, Nd^{3+}, Sm^{3+}, Eu^{3+}, Gd^{3+}, and Er^{3+}. The ionic radii of lanthanide ions are presumed to play an important role in the synthesis. Only Ln^{3+} ions of larger ionic radii can adopt such 12-coordinated icosahedral configuration. For those lanthanide ions following Pr^{3+}, only $[LnNi_6]$ trigonal prismatic clusters have been obtained [122]. Comparing with the octahedral $[SmNi_6]$ cluster synthesized in acetonitrile [66], we can see that the solvent can also affect the synthesis of the octahedral cluster, as the coordination ability of water to Ln^{3+} is superior to a nonaqueous solvent molecule. The competition of solvent and ionic radii effects account for the fact that $[LnNi_6]$ (Ln = Nd–Lu) octahedral clusters cannot be synthesized in aqueous solution. The $[LnNi_6]$ octahedral cluster is very stable and does not take up any additional ligand, such as imidazole or SCN^- [121].

Upon cooling, the $\chi_M T$ value for **16** increases slightly when the temperature is cooled to about 60 K and more rapidly up to about 7 K, when it reaches a maximum of $9.79 \, cm^3 \, mol^{-1} \, K$. This phenomenon corresponds to a ferromagnetic interaction. Further decrease of $\chi_M T$ below 7 K may be due to anisotropic effects, like zero-field splitting or spin–orbit coupling. This Ni–Ni ferromagnetic interaction in the $[LaNi_6]$ cluster is also confirmed by the positive Weiss constants (2.35 K) determined with data collected in the temperature range 10–300 K.

8.2 $[LnM_6]$ Trigonal Prismatic Clusters ($M = Co^{2+}, Ni^{2+},$ and Zn^{2+})

With glycine as ligand, a series of heptanuclear trigonal–prismatic polyhedra $[LnM_6]$ (M = Co, Ni, Zn) with different edge and terminal ligands can be obtained (Fig. 18) [122, 125]. The common structure of these clusters may be described as a 4f metal ion at the center of a large trigonal prism formed by six 3d metal ions. The Ni clusters are here taken as illustrative examples. The complexes all have a $[LnNi_6(gly)_6(\mu_3-OH)_3(H_2O)_6]^{6+}$ core and can be viewed as the edge ligand and/or terminal ligand exchange products of a primordial species $[LnNi_6(gly)_6(\mu_3-OH)_3(H_2O)_6(\mu_2-OH_2)_3]^{6+}$ (unfortunately, this parent cluster has not been synthesized). Actually, the μ_2-H_2O edge ligands with Ni–O distance of ca. 2.23 Å and the terminal H_2O (Ni–O 2.17 Å) groups coordinate to nickel ions very

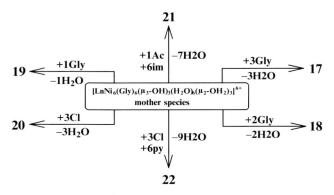

Scheme 4 Formation of the heptanuclear trigonal prismatic clusters **17–22**

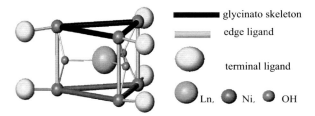

Fig. 18 Heptanuclear trigonal prismatic clusters with amino acids as ligands

weakly, as compared with the normal Ni–O bond length of ca. 2.0 Å. The μ_2–H$_2$O edge groups can be replaced with a Cl$^-$ or Ac$^-$ group one by one, while the terminal ones can be substituted by pyridine. Besides the alteration of lanthanides, a series of heterometal clusters can be obtained. Scheme 4 shows the formation of the clusters, and their structures (or possible structure) are shown in Fig. 18.

For example, if the three μ_2–OH$_2$ edge ligands are replaced by three, two, or one glycinato ligands in [2.110] coordination mode, Na$_4$[PrNi$_6$(gly)$_9$(μ_3–OH)$_3$(H$_2$O)$_6$] · 7ClO$_4$ (**17**), Na$_2$[PrNi$_6$(gly)$_8$(μ_3–OH)$_3$(μ_2–OH$_2$)(H$_2$O)$_6$] · 6ClO$_4$ · 2H$_2$O (**18**), and Na[DyNi$_6$(gly)$_7$(μ_3–OH)$_3$(μ_2–OH$_2$)$_2$(H$_2$O)$_6$] · 6ClO$_4$ · H$_2$O (**19**) can be obtained (Fig. 19). The cluster in **17** contains two parallel layers, each being composed of three Ni^{2+} ions and three glycinato ligands. The Ni···Ni distances are about 3.6 Å (different layers) and 5.3 Å (same layer) respectively, while the Pr···Ni distance is about 3.57 Å. The three glycinato ligands adopt the [3.1$_1$2$_23$1$_3$] coordination mode, chelating to two Ni^{2+} and one Ln^{3+} ion. Pr^{3+} has a nine-coordinated tricapped trigonal–prismatic coordination polyhedron. The structures of **18** and **19** are almost the same as that of **17** except for the number of edge glycinato ligands. This also causes a slight distortion of the metal skeleton. In other words, a 4.5° torsion angle may be formed between the two layers in **18** and **19**. The three μ_2–OH$_2$ edge ligands can also be replaced partly or wholly by Cl$^-$ to form new species such as ([SmNi$_6$(gly)$_6$(μ_3–OH)$_3$Cl$_3$(H$_2$O)$_6$] · 3Cl · 9H$_2$O (**20**).

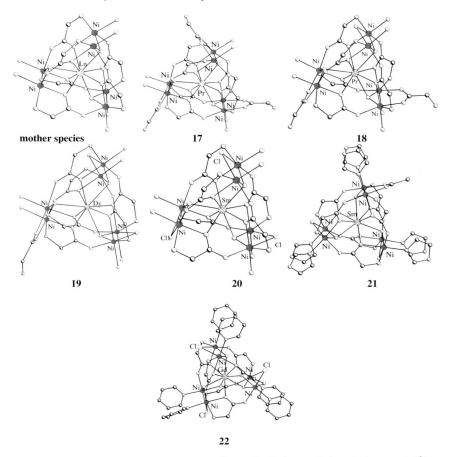

Fig. 19 Possible structure of the mother species [LnNi$_6$(gly)$_6$(μ_3–OH)$_3$(H$_2$O)$_6$(μ_2–OH$_2$)$_3$]$^{6+}$ and the structures of some heptanuclear trigonal prismatic clusters **17–22**

The six terminal H$_2$O ligands of the primordial species [LnNi$_6$(gly)$_6$(μ_3–OH)$_3$(H$_2$O)$_6$(H$_2$O)$_6$]$^{6+}$ can be substituted by imidazole (im) or pyridine (py) ligands, which leads to ([SmNi$_6$(gly)$_6$(Ac)(μ_3–OH)$_3$(μ_2–H$_2$O)$_2$(im)$_6$] · (ClO$_4$)$_5$ · (H$_2$O)$_5$) (**21**) and ([GdNi$_6$(gly)$_6$Cl$_3$(OH)$_3$(py)$_6$] · Cl · (ClO$_4$)$_2$ · (H$_2$O)$_5$) (**22**), respectively [125]. This is reminiscent of other well-known polynuclear clusters whose terminal sites are occupied by solvent molecules and can be substituted by pyridine, carboxylic acid, or another ligand, such as trinuclear clusters [M$_3$(μ_3–O)(μ–RCO$_2$)$_6$L$_3$]$^{n+}$ (L = neutral monodentate ligand) widely occurring in a variety of first-, second-, or third-row transition metal ions [126], and polynuclear manganese clusters [127, 128]. However, similar reaction is still very rare in 3d–4f heteronuclear clusters. Similar [LnCu$_6$] clusters [LnCu$_6$(μ_3–OH)$_3$(gly)$_6$(im)$_6$](ClO$_4$)$_6$ (Ln = La, Pr, Sm, Er) (**1**) have also been synthesized [79].

The replacement of groups with weak coordination interaction, including substitution of the edge and terminal ligands, plus the alteration of lanthanides, may be of help in the design and synthesis of a large variety of 3d–4f clusters. This approach is attractive and it may open up new routes for the construction of other polynuclear complexes. A similar situation also applies to analogous Zn^{2+} and Co^{2+} clusters.

It is worth noting that the fragment $[Ni_2(gly)_2(OH)(Hgly)]^+$, similar to metalloligand c in Scheme 2, exists in compounds **17–19**. Thus, an interesting question arises as to why the fragment $[Ni_2(gly)_2(OH)(Hgly)]^+$ chelates the lanthanide ions rather than an octahedral divalent transition metal ion. Firstly, the octahedral cluster $[Ln_6Ni_{12}(OH)_{24}(ClO_4)]^{17+}$ is still not observed. On the other hand, due to the octahedral coordination environment of Ni^{2+} ions, the Hgly only replaces the weakly coordinated μ_2–H_2O edge ligand, resulting in the large Ni–Ni distance of ca. 3.5 Å and the small torsion angle of the two Ni(gly) moieties (about 64.2°). Therefore, the $[Ni_2(gly)_2(OH)(Hgly)]^+$ fragment, similar to the metalloligand $[Cu_2(gly)(OH)(im)_2]^{2+}$ (*h* in Scheme 3), chelates the lanthanides to form the trigonal prismatic cluster.

9 Other Factors Affecting the Assembly

Except for the factors mentioned above, such as the reactant ratio employed, variation of lanthanide and transition metal, crystallization conditions, and the presence of a secondary ligand, there are several other factors that can affect the controllable assembly of the lanthanide–transition metal–amino acid cluster compounds.

Firstly, the difficulty in this strategy is seeking a suitable pH window so that the amino acid ligand can use both its amino and carboxylate groups to coordinate to 3d and 4f metal ions simultaneously. We found that 6.6 is a suitable pH, and that higher pH generally leads to a large amount of precipitate.

Another key factor is the choice of reactant proportion. We found that it is necessary to maintain a high metal-to-ligand ratio because a low ratio (such as $Ln:Cu:gly:Ac^- = 1:4:2:2$) will lead to the final product being contaminated by a small number of light-blue crystals of $Cu_3(gly)_4(ClO_4)_2$.

Finally, the solubility of the amino acids is also important. Only those amino acids that are readily soluble in water, such as glycine, alanine, valine, and proline, can be employed to construct the present family of 3d–4f amino acid compounds.

10 Conclusions

We have summarized our research findings in a systematic study of lanthanide–transition metal–amino acid clusters. Several factors of influence that affect the assembly, such as the presence of a secondary ligand, variation of lanthanide, crystallization conditions, the ratio of metal ions to amino acids, and the choice of

Controllable Assembly, Structures, and Properties of Lanthanide–Transition   203

transition metal ions have been expounded. The substitution of weak coordination bonds is identified as an important reason for the diverse structures of this series of compounds. The diverse structures found for the Cu–Ln–gly clusters may be attributed to the dynamic balance of several metalloligands and the versatile coordination modes of the Cu^{2+} ion.

Acknowledgement This work was supported by the grants from NNSF of China (20673118 and 20733003), NSF of Fujian Province (2005HZ01-1 and 2006J0014), 973 program (2007CB815301 and 2006CB932900) and the Chinese Academy of Sciences (KJCX2-YW-M05).

References

1. Fenske D, Anson CE, Eichhöfer A, Fuhr O, Ingendoh A, Persau C, Richert C (2005) Angew Chem Int Ed 44:5242–5246
2. Tasiopoulos AJ, Vinslava A, Wernsdorfer W, Abboud KA, Christou G (2004) Angew Chem Int Ed 43:2117–2121
3. Müller A, Beckmann E, Bögge H, Schmidtmann M, Dress A (2002) Angew Chem Int Ed 41:1162–1167
4. Tran NT, Powell DR, Dahl LF (2000) Angew Chem Int Ed 39:4121–4125
5. Krautscheid H, Fenske D, Baum G, Semmelmann M (1993) Angew Chem Int Ed Engl 32:1303–1306
6. Wu XT (2005) Inorganic assembly chemistry. Science, Beijing, China
7. Zhu NY, Du SW, Wu XT, Lu JX (1992) Angew Chem Int Ed Engl 31:87–88
8. Du SM, Zhu NY, Chen PC, Wu, XT (1992) Angew Chem Int Ed Engl 31:1085–1087
9. Huang Q, Wu XT, Wang QM, Sheng TL, Lu JX (1996) Angew Chem Int Ed 35:868–870
10. Guo J, Wu XT, Zhang WJ, Sheng TL, Huang Q, Lin P, Wang QM, Lu JX (1997) Angew Chem Int Ed 36:2464–2466
11. Yu H, Zhang WJ, Wu XT, Sheng TL, Wang QM, Lin P (1998) Angew Chem Int Ed 37:2520–2522
12. Heo J, Jeon YM, Mirkin CA (2007) J Am Chem Soc 129:7712–7713
13. Zhao SB, Wang RY, Wang S (2007) J Am Chem Soc 129:3092–3093
14. Yoshizawa M, Tamura M, Fujita M (2006) Science 312:251–254
15. Fiedler D, Leung DH, Bergman RG, Raymond KN (2004) J Am Chem Soc 126:3674–3675
16. Perry JJ, Kravtsov VC, McManus GJ, Zaworotko MJ (2007) J Am Chem Soc 129:10076–10077
17. Sudik AC, Millward AR, Ockwig NW, Cote AP, Kim J, Yaghi OM (2005) J Am Chem Soc 127:7110–7118
18. Eddaoudi M, Kim J, Wachter JB, Chae HK, O'Keeffe M, Yaghi OM (2001) J Am Chem Soc 123:4368–4369
19. Xiang SC, Wu XT, Zhang JJ, Fu RB, Hu SM, Zhang XD (2005) J Am Chem Soc 127:16352–16353
20. Serre C, Mellot-Draznieks C, Surblé S, Audebrand N, Filinchuk Y, Férey G (2007) Science 315:1828–1831
21. Maji TK, Matsuda R, Kitagawa S (2007) Nat Mater 6:142–148
22. Férey G, Mellot-Draznieks C, Serre C, Millange F, Dutour J, Surblé S, Margiolaki I (2005) Science 309:2040–2042
23. Matsuda R, Kitaura R, Kitagawa S, Kubota Y, Belosludov RV, Kobayashi TC, Sakamoto H, Chiba T, Takata M, Kawazoe Y, Mita Y (2005) Nature 436:238–241
24. Chae HK, Siberio-Pérez DY, Kim J, Go Y, Eddaoudi M, Matzger AJ, O'Keeffe M, Yaghi OM (2004) Nature 427:523–527

25. Aronica C, Pilet G, Chastanet G, Wernsdorfer W, Jacquot JF, Luneau D (2006) Angew Chem Int Ed 45:4659–4662
26. Zaleski CM, Depperman EC, Kampf JW, Kirk ML, Pecoraro VL (2004) Angew Chem Int Ed 43:3912–3914
27. Tasiopoulos AJ, O'Brien TA, Abboud KA, Christou G (2004) Angew Chem Int Ed 43:345–349
28. Blake AJ, Milne PEY, Winpenny REP, Thornton P (1991) Angew Chem Int Ed Engl 30:1139–1141
29. Mereacre VM, Ako AM, Clerac R, Wernsdorfer W, Filoti G, Bartolome J, Anson CE, Powell AK (2007) J Am Chem Soc 129:9248–9249
30. Ferbinteanu M, Kajiwara T, Choi KY, Nojiri H, Nakamoto A, Kojima N, Cimpoesu F, Fujimura Y, Takaishi S, Yamashita M (2006) J Am Chem Soc 128:9008–9009
31. Mishra A, Wernsdorfer W, Abboud KA, Christou G (2004) J Am Chem Soc 126:15648–15649
32. Costes JP, Dahan F, Wernsdorfer W (2006) Inorg Chem 45:5–7
33. Mori F, Nyui T, Ishida T, Nogami T, Choi KY, Nojiri H (2006) J Am Chem Soc 128:1440–1441
34. Murugesu M, Mishra A, Wernsdorfer W, Abboud KA, Christou G (2006) Polyhedron 25:613–625
35. Chen QY, Luo QH, Hu XL, Shen MC, Chen JT (2002) Chem Eur J 8:3984–3990
36. Gunnlaugsson T, Leonard JP, Senechal K, Harte AJ (2004) Chem Commun, pp 782–783
37. Zhao B, Chen XY, Chen P, Liao DZ, Yan SP, Jiang ZH (2004) J Am Chem Soc 126:15394–15395
38. Zhao B, Gao HL, Chen XY, Cheng P, Shi W, Liao DZ, Yan SP, Jiang, ZH (2006) Chem Eur J 12:149–158
39. Skaribas SP, Pomonis PJ, Sdoukos AT (1991) J Mater Chem 1:781–784
40. Hasegawa E, Aono H, Igoshi T, Sakamoto M, Traversa E, Sadaoka Y (1999) J Alloys Compd 287:150–158
41. Winpenny REP (1998) Chem Soc Rev 27:447–452
42. Sakamoto M, Manseki K, Okawa H (2001) Coord Chem Rev 219–221:379–414
43. Sakagami N, Okamoto K (1998) Chem Lett 27:201–202
44. Decurtins S, Gross M, Schmalle HW, Ferlay S (1998) Inorg Chem 37:2443–2449
45. Cutland AD, Malkani RG, Kampf JW, Pecoraro VL (2000) Angew Chem Int Ed 39:2689–2691
46. Kahn ML, Verelst M, Lecantes M, Mathoniere C, Kahn O (1999) Eur J Inorg Chem, pp 527–531
47. Sanada T, Suzuki T, Kaizaki S (1998) J Chem Soc Dalton Trans, pp 959–965
48. Plecnik CE, Liu SM, Shore SG (2003) Acc Chem Res 36:499–508
49. Chen XM, Yang YY (2000) Chin J Chem 18:664–672
50. Wang RY, Gao F, Jin TZ (1996) Huaxuetongbao 10:14–20
51. Ohata N, Masuda H, Yamauchi O (1996) Angew Chem Int Ed 35:531–532
52. Zheng ZP (2001) Chem Commun, pp 2521–2529
53. Hu SM, Du WX, Dai JC, Wu LM, Cui CP, Fu ZY, Wu XT (2001) J Chem Soc Dalton Trans, pp 2963–2964
54. Wang LY, Igarashi S, Yukawa Y, Hoshino Y, Roubeau O, Aromí G. Winpenny REP (2003) J Chem Soc Dalton Trans, pp 2318–2324
55. Du M, Bu XH, Guo YM, Ribas J (2004) Chem Eur J 10:1345–1354
56. Xiang SC, Hu SM, Zhang JJ, Wu XT, Li JQ (2005) Eur J Inorg Chem 2706–2714
57. Ama T, Rashid MM, Saker AK, Miyakawa H, Yonemura T, Kawaguchi H, Yasui T (2001) Bull Chem Soc Jpn 74:2327–2333
58. Okamoto KI, Aizawa SI, Konno T, Einaga H, Hidaka J (1986) Bull Chem Soc Jpn 59:3859–3864
59. Igashira-Kamiyama A, Fujioka J, Kodama T, Kawamoto T, Konno T (2006) Chem Lett 35:522–523

Controllable Assembly, Structures, and Properties of Lanthanide–Transition 205

60. Strasdeit H, Busching I, Behrends S, Saak W, Barklage W (2001) Chem Eur J 7:1133–1142
61. Abu-Nawwas AH, Cano J, Christian P, Mallah T, Rajaraman G, Teat SJ, Winpenny REP, Yukawa Y (2004) Chem Commun, pp 314–315
62. Anokhina EV, Jacobson AJ (2004) J Am Chem Soc 126:3044–3045
63. Coxall RA, Harris SG, Henderson DK, Parsons S, Tasker PA, Winpenny REP (2000) J Chem Soc Dalton Trans, pp 2349–2356
64. Anokhina EV, Go YB, Lee Y, Vogt T, Jacobson AJ (2006) J Am Chem Soc 128:9957–9962
65. Vaidhyanathan R, Bradshaw D, Rebilly JN, Barrio JP, Gould JA, Berry NG, Rosseinsky MJ (2006) Angew Chem Inter Ed 45:6495–6499
66. Yukawa Y, Igarashi S, Yamano A, Sato S (1997) Chem Commun, pp 711–712
67. Igarashi S, Hoshino Y, Masuda Y, Yukawa Y (2000) Inorg Chem 39:2509–2515
68. Yukawa Y, Aromí G, Igarashi S, Ribas J, Zvyagin SA, Krzystek J (2005) Angew Chem Int Ed 44:1997–2001
69. Gao F, Wang RY, Jin TZ, Xu GX, Zhou ZY, Zhou XG (1997) Polyhedron 16:1357–1360
70. Li ZS, Sun HL, Kou HZ, Han ST, Gao S (2002) J Rare Earth 20:343–347
71. Yamaguchi T, Sunatsuki Y, Kojima M, Akashi H, Tsuchimoto M, Re N, Osa S, Matsumoto N (2004) Chem Commun, pp 1048–1049
72. Casellato U, Guerriero P, Tamburini S, Sitran S, Vigato PA (1991) J Chem Soc Dalton Trans, pp 2145–2152
73. Costes J-P, Dahan F, Dumestre F, Clemente-Juan JM, Garcia-Tojal J, Tuchagues J-P (2003) Dalton Trans, pp 464–468
74. Hu, SM, Xiang SC, Zhang JJ, Sheng TL, Fu RB, Wu XT (2008) Eur J Inorg Chem, pp 1141–1146
75. Fu ZY, Wu XT, Dai JC, Wu LM, Cui CP, Hu SM (2001) Chem Commun, pp 856–1857
76. Ako AM, Hewitt IJ, Mereacre V, Clérac R, Wernsdorfer W, Anson CE, Powell, AK Angew Chem Int Ed 45:4926–4929
77. Murugesu M, Clérac R, Anson CE, Powell AK (2004) Chem Commun, pp 598–1599
78. Murugesu M, Clérac R, Anson CE, Powell AK (2004) Inorg Chem 43:7269–7271
79. Du, WX, Zhang JJ, Hu SM, Xia SQ, Fu RB, Xiang SC, Li YM, Wang LS, Wu XT (2004) J Mol Struct 701:25–30
80. Zhang JJ, Sheng TL, Xia SQ, Leibeling G, Meyer F, Hu SM, Fu RB, Xiang SC, Wu XT (2004) Inorg Chem 43:5472–5478
81. Chen XM, Aubin SMJ, Wu YL, Yang YS, Mak TCW, Hendrickson DN (1995) J Am Chem Soc 117:9600–9601
82. Chen XM, Wu YL, Tong YX, Huang XY (1996) J Chem Soc Dalton Trans, pp 2443–2448
83. Yang YY, Chen XM, Ng SW (2001) J Solid State Chem 161:214–224
84. Yang YY, Huang ZQ, He F, Chen XM, Ng SW (2004) Z Anorg Allg Chem 630:286–290
85. Cui Y, Chen JT, Huang JS (1999) Inorg Chim Acta 293:129–139
86. Boyd PDW, Li Q, Vincent JB, Folting K, Chang HR, Streib WE, Huffmann JC, Christou G, Hendrickson DN (1988) J Am Chem Soc 110:8537–8539
87. Soler M, Wemsdorfer W, Folting K, Pink M, Christou G (2004) J Am Chem Soc 126:2156–2165
88. Crawford VH, Richardson HW, Wasson JR, Hodgson DJ, Hatfield WE (1976) Inorg Chem 15:2107–2110
89. Figgis BN, Hitchman MA (2000) Ligand field theory and its applications. Wiley, Toronto, chaps. 9 and 11
90. Coronado E, Day P (2004) Chem Rev 104:5419–5448
91. Zhang ZJ, Xiang SC, Zhang YF, Wu AQ, Cai LZ, Guo GC, Huang JS (2006) Inorg Chem 45:1972–1977
92. Zhang JJ, Hu SM, Xiang SC, Sheng TL, Wu XT, Li YM (2006) Inorg Chem 45:7173–7181
93. Hu SM, Dai JC, Wu XT, Wu LM, Cui CP, Fu ZY, Hong MC, Liang YC (2002) J Cluster Sci 13:33–41
94. Zhang JJ, Xia SQ, Sheng TL, Hu SM, Leibeling G, Meyer F, Wu XT, Xiang SC, Fu RB (2004) Chem Commun, pp 1186–1187

95. Zhang JJ, Sheng TL, Hu SM, Xia SQ, Leibeling G, Meyer F, Fu ZY, Chen L, Fu RB, Wu XT (2004) Chem Eur J 10:3963–3969
96. Blake AJ, Gould RO, Grant CM, Milne PEY, Parsons S, Winpenny REP (1997) J Chem Soc Dalton Trans, pp 485–496
97. Zheng NF, Bu XH, Feng PY (2002) J Am Chem Soc 124:9688–9689
98. Rodriguez-Fortea A, Alemany P, Alvarez S, Ruiz E (2001) Chem Eur J 7:627–637
99. Férey G (2003) Angew Chem Int Ed 42:2576–2579
100. Xiang SC, Hu SM, Sheng TL, Fu RB, Wu XT, Zhang XD (2007) J Am Chem Soc 129:15144–15146
101. Dalgarno SJ, Raston CL (2003) Dalton Trans, pp 287–290
102. Atwood JL, Barbour LJ, Dalgarno S, Raston CL, Webb HR (2002) J Chem Soc Dalton Trans, pp 4351–4356
103. Brügstein MR, Gamer MT, Roesky PW (2004) J Am Chem Soc 126:5213–5218
104. Wang R, Song D, Wang S (2002) Chem Commun, pp 368–369
105. Brügstein MR, Roesky PW (2000) Angew Chem Int Ed 39:549–551
106. Xu J, Raymond KN (2000) Angew Chem Int Ed 39:2745–2747
107. Mudring AV, Timofte T, Babai A (2006) Inorg Chem 45:5162–5166
108. Fang X, Anderson TM, Benelli C, Hill CL (2005) Chem Eur J 11:712–718
109. Zhang DS, Ma BQ, Jin TZ, Gao S, Yan CH, Mak TCW (2000) New J Chem 24:61–62
110. Wang R, Carducci MD, Zheng Z (2000) Inorg Chem 39:1836–1837
111. Žák Z, Unfried P, Giester G (1994) J Alloys Compd 205:235–242
112. Panagiotopoulos A, Zafiropoulos TF, Perlepes SP, Bakalbassis E, Massonramade I, Kahn O, Terzis A, Raptopoulou CP (1995) Inorg Chem 34:4918–4920
113. Freedman DE, Bennett MV, Long JR (2006) Dalton Trans, pp 2829–2834
114. Zhong ZJ, Seino H, Mizobe Y, Hidai M, Fujishima A, Ohkoashi S, Hashimoto K (2000) J Am Chem Soc 122:2952–2953
115. Hatscher ST, Urland W (2003) Angew Chem Int Ed 42:2862–2864
116. Costes JP, Clemente-Juan JM, Dahan F, Nicodème F, Verelst M (2002) Angew Chem Int Ed 41:323–325
117. Costes JP, Dahan F, Dupuis A (2000) Inorg Chem 39:165–168
118. Freeman HC, Snow MR, Nitta I, Tomita K (1964) Acta Crystallogr 17:1463–1470
119. Friedrichs OD, O'Keeffe M, Yaghi OM (2003) Acta Crystallogr A 59:515–525
120. Delf BW, Gillard RD, O'Brien P (1979) J Chem Soc Dalton Trans, pp 1301–1305
121. Zhang JJ, Xiang SC, Hu SM, Xia SQ, Fu RB, Wu XT, Li YM, Zhang HS (2004) Polyhedron 23:2265–2272
122. Zhang JJ, Hu SM, Zheng LM, Wu XT, Fu ZY, Dai JC, Du WX, Zhang HH, Sun RQ (2002) Chem Eur J 8:5742–5749
123. Zhang JJ, Hu SM, Xiang SC, Wang LS, Li YM, Zhang HS, Wu XT (2005) J Mol Struct 748:129–136
124. Doble DMJ, Benison CH, Blake AJ, Fenska D, Jackson MS, Kay RD, Li WS, Schroder M (1999) Angew Chem Int Ed 38:1915–1918
125. Zhang JJ, Hu SM, Xiang SC, Wu XT, Wang LS, Li YM (2006) Polyhedron 25:1–8
126. Abe M, Sasaki Y, Yamada Y, Tsukahara K, Yano S, Ito T (1995) Inorg Chem 34:4490–4998
127. Baikie ARE, Howers AJ, Hursthouse MB, Quick AB, Thornton P (1986) J Chem Soc Chem Commun, pp 1587–1588
128. Low DW, Eichhorn DM, Draganescu A, Armstrong WH (1991) Inorg Chem 30:877–878

Index

Acetate 93
Acetate ligand, bidentate 169
Al_6Mn 2
Alaninato ligands 164
Amino acids 161
   coordination chemistry/binding modes 163
Angle-resolved XPS (ARXPS) 48, 86
Antimonides 41, 79
Approximant crystals 1
   Bergman-type 5, 32
   Mackay-type 5
   Tsai-type 5, 27
Arsenide phosphides 72
   mixed 78
Arsenides 41, 72
   transition-metal 72
Aspartate ligand 163
Auger electron spectroscopy (AES) 51
Auger emission 51
Azobis(5-chloropyrimdine) (abcp) 116
Azobispyridine (abpy) 116
Bergman cluster 5
4,4′-Bipyridine 166
Bis[2-(diphenylphosphino)ethyl]
   phenylphosphine 106
Ca–Au–Ga 25
Ca–Au–In 25
Chalcogenides 8
Cluster compounds, 3d–4f 161
Cluster-to-ligand charge transfer 108
$CoAs_3$ 81
CoP_3 81
Concentric hemispherical analyser (CHA) 46
Copper clusters, polynuclear 163
Core line spectra 44
$CoSb_3$ 81
Cu–Ln–gly compounds 193, 197

Cyanopyridine 98, 100, 124
Diglycinato–copper 186
1,3-Diiminosioindoline 123
N,N-Dimethylaminopyridine 104
Electron phases 7
Electron spectroscopy for chemical analysis
   (ESCA) 45
Electronic interactions 93
Electronic tuning 1
Er:Cu:gly 190
Extended X-ray absorption fine structure
   (EXAFS) 59
Fibonacci sequence 4
Galium 16
Gd:Cu:gly 181
Glue atoms 27
Glycinato ligands 177, 192, 195, 197, 200
Halides 8
Heptanuclear trigonal prismatic clusters 200
Hexaimine ligands 109
$Hf(Si_{0.5}As_{0.5})As$ 54, 86
Hume–Rothery rules 7
Icosahedral quasiperiodic crystals (i-QC) 2
Imidazole ligand, monodentate 166
Imidazole/acetate, 3d–4f clusters 166
In-plane coordination pattern 124
Indium 16
Inelastic mean free path (IMFP) 48
Intermetallics 1
   polar 9
Intervalence charge transfer (IVCT) 99
Intracluster transition 99
Lanthanide–copper–glycinato coordination
   polymers 187, 194
Lathanide(III)–transition metal–amino acid
   161
Layer-by-layer 98

207

Ligand substitution 93
  reactivity 95
Ligands, ditopic N-heterocyclic 99
Mackay cluster 5
N-Methyl-4,4′-bipyridiniumion 97
$Mg_2Cu_6Ga_5$ 16
Mixed arsenide phosphides 72
Monopnictides 79
Multiplet splitting 52
Ni aspartate (asp) coordination polymers 163
Out-of-plane coordination pattern 124
Oxides 8
Phosphides 41
  mixed-metal 68
  MnP-type binary 72
  transition-metal 64
Phosphorus 2p XPS 69
Photoemission spectroscopy (PES) 49, 58
Phthalic acid 123
Phthalic anhydride 123
Phthalocyanines 121
  catalytic properties 157
Phthalocyanines, conductivity 153
  crystal structures 125
  ionic 147
  magnetic properties 156
  neutral parent 126
  neutral parent in-plane, axial ligands 128
  neutral parent out-of-plane 134
  peripherally symmetrical substituted 138
  peripherally unsymmetrical substituted 146
  synthesis/modifications 123
  UV–vis spectra 151
Phthalonitrile 123
Phthalyl derivatives, cyclotetramerization 123
Phthlimide 123
Plasmon loss 53
Pnictides, MnP-type structure 61
Polyanionic clusters, Zintl phases 1
Prolate rhombohedron 28
Pseudogap 1
  tuning 16
Pyrazine 99
Pyrazino[2,3-f]quinoxaline 115
Quasicrystals 1
  approximants 10
  Mg_2Zn_{11}-type precursors 18
  modeling 34

Ruthenium–carboxylate cluster, oxo-centered
  trinuclear 94
Sc–Mg–Zn 20, 24
Sc_3Zn_{17} 14
Skutterudites 79
Spin–orbit coupling 51
Structure regularities 27
Sulfide systems 8
Synchrotron radiation (SR) 48
Take-off angle 48
Tetraimine ligands 109
Traditional metal clusters 8
Transition-metal arsenides 72
Transition-metal phosphides 64
Triacontanuclear octahedral clusters, bidentate
  acetate ligand 167
Triels, polar intermetallics 10
Triphenylphosphine 102
Triruthenium 93
  methanol-coordinated 108
Triruthenium complexes, axial ligand
  substitution 95
  azo–aromatic ligand-substituted 116
  bridging ligand substitution 108
  N-heterocyclic ligand-substituted 96
  ortho-metallated ligands 108
  phosphine ligand-substituted 102
Triruthenium-acetate cluster, oxo-centered 93
Tsai clusters 5, 27
Ultraviolet photoelectron spectroscopy (UPS)
  53
Valence band spectra 44
X-ray absorption near-edge spectroscopy
  (XANES) 58
  K-edge 69
X-ray absorption spectroscopy 58
  arsenides/phosphides 41
X-ray fluorescence (XRF) 61
X-ray photoelectron spectroscopy (XPS) 44
  arsenides/phosphides 41
  instrumentation 45
  metal 2p 70
  phosphorus 2p 69, 74
  spectra, core level 49
  valence band spectra 53
X-ray spectroscopies 43
Zintl boundary 10
Zintl phases 8